INTERNET POLICIES AND ISSUES

INTERNET ISSUES: BLOGGING, THE DIGITAL DIVIDE AND DIGITAL LIBRARIES

INTERNET POLICIES AND ISSUES

Additional books in this series can be found on Nova's website under the Series tab.

Additional E-books in this series can be found on Nova's website under the E-book tab.

INTERNET POLICIES AND ISSUES

INTERNET ISSUES: BLOGGING, THE DIGITAL DIVIDE AND DIGITAL LIBRARIES

CASSIE M. EVANS
EDITOR

Nova
Nova Science Publishers, Inc.
New York

For permission to use material from this book please contact us:
Telephone 631-231-7269; Fax 631-231-8175
Web Site: http://www.novapublishers.com

LIBRARY OF CONGRESS CATALOGING-IN-PUBLICATION DATA

Internet issues : blogging, the digital divide and digital libraries /
editor, Cassie M. Evans.
p. cm.
Includes bibliographical references and index.
ISBN 978-1-60741-824-5 (hbk)
1. Internet--Social aspects. 2. Blogs. 3. Digital divide. 4. Digital
libraries. I. Evans, Cassie M.
HM851.I5824 2009
302.23'1--dc22

2010002305

Published by Nova Science Publishers, Inc. ✦ New York

CONTENTS

Preface vii

Chapter 1 The Science of Blogging: About Science 1
Christine Redman

Chapter 2 A Hyperlink Study of Greek Political Blogs Communication Patterns 27
Kostas Zafiropoulos and Vasiliki Vrana

Chapter 3 Understanding the Academic Blogosphere: A Look into the Generic Tangle 53
María José Luzón

Chapter 4 Blogging in Higher Education 71
Patricia D. Wolf and Richard W. Smith

Chapter 5 Librarians and Information Scientists Blogosphere: An Overview 83
Noa Aharony

Chapter 6 The Data Deluge: Opportunities for Research in Educational Digital Libraries 95
Beijie Xu, Mimi Recker and Sherry Hsi

Chapter 7 The Role of Family on Increasing Participation and Bridging the Digital Divide: Evidence from an E-Health Study 117
Sally Lindsay, Simon Smith and Paul Bellaby

Chapter 8 Digital Outdistancing: Socio-Economic Status Differences in Use of the Internet, What Could It Mean for Health Disparities? 131
Silje C. Wangberg

Chapter 9 Shut Up And Listen: The Influence of Selective Exposure to Blogs on Political Tolerance 147
Thomas J. Johnson, Shannon L. Bichard and Weiwu Zhang

Chapter 10 Equity, Justice and the Digital Divide in a Pandemic 165
Connal Lee

Chapter 11 Understanding "Serious Videogame" Storyline and Genre Preferences Related To Game Immersion Among Low-Income Ethnically Diverse Urban And Rural Adolescents **177**
Victoria Thompson, Debbe Thompson and Tom Baranowski

Chapter 12 Increasing Unstructured Information Reuse **189**
Rogerio Atem de Carvalho, Ronaldo Amaral Santos and Vanderson Mota dos Santos

Chapter 13 Twitter: A Productive Learning Tool for the SMS Generation **199**
Adelina Moura and Ana Amélia Carvalho

Expert Commentary
Disability and the Digital Divide: Gaps and Future Directions **215**
Sally Lindsay

Index **221**

PREFACE

A blog is a type of website, usually maintained by an individual with regular entries of commentary, descriptions of events, or other material such as graphics or video. The term digital divide refers to the gap between people with effective access to digital and information technology and those with very limited or no access at all. It includes the imbalances in physical access to technology as well as the imbalances in resources and skills needed to effectively participate as a digital citizen. A digital library is a library in which collections are stored in digital formats. This book provides insight into the technologies provided by the internet such as blogging, digital libraries and the digital divide, which both enhance communication between certain individuals and limit others.

Chapter 1 - This chapter has three main components. It provides a theoretical argument for blogging, a structure for introducing blogging into the educational landscape and a methodology to objectively code the blogging material as useful research data. The theoretical premise behind blogging in education for learning is based on the Vygotskian approach that values learning from peers in a dialectical structure. This chapter comprehensively explores the theoretical arguments underpinning the use of blogging for learning. It seeks to demonstrate how the theory informs a pedagogical structure that educators could implement when introducing blogging as and for a science learning experience. The pedagogical structure has the potential to support educators to assist the introduction as well as the progression and development of the blogging experience for learners. This focus on blogging is part of a broader perspective that Information Communication technology may be able to contribute to enhanced learning outcomes in science classrooms, though more independent engagement and the use of critical thinking skills. By utilising a socio-cultural theory of learning for the data analysis, the blogging material can be viewed as revealing the ontology of the writers. The chapter examines the contribution of blogging as a function of new technologies that may assist the enrichment of learning outcomes in science education in ways that better inform classroom practices in practical ways.

Chapter 2 - Blogs are web pages that accept comments-posts of users in the form of a diary. They allow posting and exchange of opinions and are part of WEB 2.0. Posting of opinions in the Internet can be considered as an expression of activated public opinion in contradiction to public opinion, which is recorded through surveys and concerns the public. The chapter starts by providing a literature review on blogging and political blogging. It presents the recent advances in the hyperlinks analysis of blogs, and describes the recent hypotheses used for this kind of research. As a case study, the chapter examines the degree of interconnectivity, cohesion and polarisation of Greek political blogs that participate in the discussion and use tags about Greek Parties. Tags are labels of thematic categories of blog posts regarding their subject. In order to cope with the research question, the study records Greek blogs with tags about parties of the Greek Parliament. Blogs with tags about Greek parties ND, PASOK, KKE, SYRIZA, and LAOS are considered. This chapter analyses connectivity patterns of Greek political blogs using a two step approach. First, it uses a Social Networking Analysis of the blogs using hyperlinks. Graph theoretic properties and indexes are used to present connectivity properties of the blogs in order to explore the fashion of blogs self-organization. The second step regards the exploration and identification of focal conversational points. The basic hypothesis supported by the literature is that within polarized political systems blogs are forming clusters around central blogs, which are considered reliable or having the same affiliation. Users of the Internet who wish to be informed quickly, locate the focal points of discussion, and for economy of navigation they read only the posts on these blogs. Bloggers also locate focal point blogs and place their posts along with a link to their blog. They expect thus that the readers of focal point blogs will also visit their blogs. The research examines whether this hypothesis holds for Greek political blogs. The quantitative analysis initially uses the theory of Social Networking. The blogs and their incoming links along with the association matrixes (0/1) of the blogs networks are recorded. Multidimensional Scaling and Hierarchical Cluster Analysis are used to locate clusters of focal point blogs. Several quantitative indexes are calculated to present the degree of interconnectivity, cohesion and polarization of blogs tagging to Greek Political Parties. Regarding the Social Network properties of the blogs, the Greek political blogosphere is connected, though not presenting high density. Regarding the formation of central blog groups, Greek political blogs conform well, though not completely, with the communication pattern supported by the literature. Study of the political and ideological affiliation of the central blog clusters, along with the study of the volume of connectivity (through hyperlinks), describes a system, which is partisan but much polarized.

Chapter 3 - Blogging is becoming increasingly popular among scholars, who see it as a helpful tool to promote their research to a broad audience and to engage in discussion about discipline-specific topics and more general (sometimes ideology-laden) issues. However, although academic weblogs share many technological and structural features, they are not a homogeneous whole, since they are written by very different types of bloggers, who use this tool for various purposes. "Academic blogging" is an umbrella term that covers a wide range of hybrid genres, customised to achieve the blogger's intended purpose and to meet the expectations of the blogging community interacting through a specific blog. Therefore, we can only discuss scholars' blogging practices if we first specify clearly the contexts of use and the social and situational factors that condition weblog discourse. The aim of this chapter is to provide an overview of academic blogs by examining the complexity of the academic blogosphere and by looking into the different types of blogs inhabiting this sphere. I draw on the faceted classification scheme for CMC proposed by Herring (2007) to analyse a sample of 10 academic blogs and account for the similarities and differences between them. This analysis will help to better comprehend academic blogs and will throw light on the factors that influence performance and communicative practices in academic blogs.

Chapter 4 - A "blog" is defined by Merriam-Webster's online dictionary as "a Web site that contains an online personal journal with reflections, comments, and often hyperlinks provided by the writer; also : the contents of such a site." According to Encyclopaedia Britannica online, "Some blogs operate mainly as news filters, collecting various online sources and adding short comments and Internet links. Other blogs concentrate on presenting original material. In addition, many blogs provide a forum to allow visitors to leave comments and interact with the publisher." "To blog" is the act of composing material for a blog. Materials are largely written, but pictures, audio, and videos are important elements of many blogs. The "blogosphere" is the online universe of blogs. Common blogging sites that offer users the ability to create blogs for free include LiveJournal, WordPress, and Google-owned Blogspot.com. Other commonly known blogs and related social websites are Tumblr, MySpace, Facebook, and LinkedIn. The microblogging site Twitter, in particular, has become a media sensation and is creating its own role in society and subsequently in education. Blogs are being used in higher education in many ways: as places faculty exchange ideas, by libraries posting news and events, as recruitment tools, as teaching tools, and by faculty to converse with students outside the classroom. However, there is dispute about the impact of blogs in higher education. This chapter reviews the ways in which blogging is being used in higher education and the debate about the impact of such uses.

Chapter 5 - Blogging, though a relatively new phenomenon, has already gained greatly in popularity, and blogs have become an essential information channel in the Web 2.0 information world. A survey conducted by the Pew/Internet American Life Project (2006) found that 8% of the 120 million internet users in the USA keep a blog, and 39% of users reported reading blogs. These data reveal an increase from a previous Pew/Internet American Life Project (2005) survey, in which only 7% of users reported writing a blog and 27% said they read blogs.

With the growing importance of blogs in our information world, the current research would like to present an overview of library and information science (LIS) blogoshere, focusing on different aspects: LIS professional literature; nature, development and prevailing tendencies in LIS blogosphere; the phenomena of tagging and folksonomies in the LIS blogosphere; and an attempt to understand comments written by readers of LIS blogs. The overview findings are important to those who are interested in blogs. It may cause them to better understand and explore the phenomenon of blogosphere and to better exploit the information assigned in the blogs.

Chapter 6 - Increasingly, educational digital libraries and associated services that support both formal and informal learning are available on the Web. As these collections are web-based, they can be engineered to capture fine-grained web usage data. However, because of their voluminous and detailed nature, rigorous analytical and interpretation strategies remain challenging. In response, *webmetrics* refers to the process of measuring, collecting, analyzing, and reporting web usage data. This chapter discusses the utility of webmetrics as a research method, and illustrates approaches for engineering and collecting web usage data in two digital library environments, the Instructional Architect and the Exploratorium's Learning Resources Collection. The chapter describes how using a set of common webmetrics such as visits, page views, and geographical location can help better understand the behaviors of end users. The authors also describe the application of *educational data mining (EDM)* to help discover meaningful patterns in usage data. Results include analyses from a pilot study that applies a clustering technique to teacher usage data from the Instructional Architect. The chapter concludes with a discussion of some of the current limitations and future possibilities of using web metrics to analyze and evaluate use and impact of educational digital libraries.

Chapter 7 – Despite the rapid increase in the diffusion of information and communication technology (ICT) the significance of older people's connection with their families on their willingness to use the Internet remains unclear. The aim of this chapter is to examine how family members helped to bridge the digital divide by influencing older people's use of the Internet. The authors' sample includes older people (n=108) from a deprived area of Greater Manchester, UK who have heart problems. Thus, they have much to gain from 'virtual mobility' because they may be less likely to have home Internet access or to visit public Internet facilities. On the other hand, since they are likely to have fewer pre-existing online social ties, their initial inclination may be lower. The authors draw on the qualitative comments from the focus groups they conducted (n=30) alongside a 9-month randomized control trial in addition to comments posted on the authors' project portal. The authors' study adds to the literature by examining the advantages of *home* computer and Internet access for those previously without them, while most studies focus on *public* use or those who already own computers. Further, by focusing on a sample that encounters multiple exclusions (e.g., deprived financially, poor health, older) we can develop a better understanding of the complex nature of the digital divide. The authors' findings suggest that grandchildren played an important part in the uptake and continued use of their computers for several people. Although some people reported that family relations had some negative influences on their Internet use, most had a positive experience and even developed new common interests with children, grandchildren, siblings and their spouses. The authors draw on Bourdieu's concept of habitus to understand the role that family had in shaping Internet use.

Chapter 8 - This chapter aims to explore relations between socio-economic status (SES), Internet use, social capital and subjective health. First, these relations are studied in the European Social Survey round 3 data set (ESS3; N = 33017). Participants were from representative samples between 15 and 80 years of age from 18 different European countries. In Study 1 it was found that SES is related to media use, in that those with a higher educational attainment and income had used the Internet more, and watched less TV. Furthermore, a simple correlation analysis showed that Internet use had positive relations to social capital variables and health, while TV has negative. When controlling for income, educational attainment, income and gender in hierarchical regression analysis, the positive relation between Internet use and subjective health remain significant but very small. The positive relation between Internet use and interpersonal trust and social interaction, respectively, were somewhat larger. A second study assessed the characteristics of users of an online smoking cessation intervention. Data on demographic characteristics of registered users in an Internet-based smoking cessation program were compared with data from an annual population smoking survey in conjunction with census data. In Study 2 it was found that the intervention had higher reach in strata of smokers that were younger, more highly educated and female. Finally, the inequalities within health online are discussed in light of the above.

Chapter 9 - The Internet has provided scholars a unique opportunity to address the issue of democracy and political polarization by exploring the relationship between political tolerance and selective exposure. This study uses an online panel of Internet users from a major university to explore the degree to which selective exposure to blogs predicts tolerance to viewpoints they oppose. Specifically, this study investigates the degree to which selective exposure as well as reliance on blogs influences political tolerance after controlling for political and demographic factors. Reliance on face-to-face discussion is also examined as a source of direct and indirect influence. The results indicate that both selective exposure and reliance on blogs have no effect on levels of political tolerance; while increased political activity, higher socioeconomic status, and reliance on face-to-face discussion appear to exert a positive influence on political tolerance. Future research should continue examining specific factors that predict political tolerance in order to better understand and preserve a democratic climate conducive to freedom of expression.

Chapter 10 - This chapter is concerned with equity and communication in a pandemic and analyses issues surrounding the potential impact of the digital divide during such an emergency. I will look at inequalities related to communication in particular access to information and how the internet can be both a positive and negative force in addressing or exacerbating these inequalities. The methodology here will be to look at the role of communication in a pandemic and establish how the internet factors into debates surrounding pandemic communication.

Communication during a pandemic is a vital consideration; ethical concerns arise regarding the possible harms posed by a pandemic and the ways in which existing inequalities in communication might place people in harms way. My argument is that the digital divide is not only potentially corrosive because access to the internet has some value but the digital divide is as much about inequalities in resources, capabilities and infrastructure. An example of this being inequalities between developed and well-resourced nations and less well-off nations and the impact that this division can have in a global pandemic. Lacking a communication infrastructure, for instance a media or technologies such as the internet is somewhat symptomatic of lacking other types of resources and structures. As such I will sketch the terrain of the issue of equity and communication and highlight the important justice concerns that surround the digital divide in a pandemic.

Chapter 11 - Various media have been embraced by researchers and non-profit organizations because they help school-aged children acquire knowledge and develop decision-making skills. Serious videogames (i.e., games designed to entertain and educate, train or change behavior) may be effective at modifying youth health behavior and their related psychosocial influences. For serious video games to be effective, however, they need to be consistent with youth game-play preferences and expectations. It is important to understand player preferences so that immersive serious games can be developed that promote desired changes. While studies have reported ethnic and gender differences in time spent playing videogames and game-play preferences, few have examined regional differences in these variables. This chapter employed both quantitative and qualitative methods (i.e., surveys and focus groups) to examine preferences for storyline genres and plot content of non-violent videogames as well as computer access, knowledge and game-play frequency in a sample of predominantly low-income urban middle school students in Texas and rural middle school students in North Carolina. Survey results revealed that White and Other/Native American male students played videogames more frequently than African-American or Hispanic male students. Although focus group participants preferred action games with some violence, they had positive reactions to challenging adventure games with male and female characters of diverse ethnicities and narratives without overt sexual overtones. Incorporating storyline and game-play preferences may promote game immersion so that the player relates to and interprets the messages based on personal experiences. Critical thinking skills undergird and complement attempts to achieve positive behavior change.

Chapter 12 - The increasing volume of electronic information available in searchable repositories has been facilitating the construction of new knowledge through information reuse and derivation. However, when it comes to unstructured information, such as text documents, retrieving whole documents, searching them "by hand", and copy-and-pasting the interesting parts is the most common form of information reuse. When it comes to more sophisticate objects, such as video and animations, the user performs a metadata search, retrieving whole potential objects. It is only by playing them that the reusable parts can be identified. The aim of this chapter is to show how it is possible - and necessary - to improve productivity to the task of unstructured information reuse by serving smaller pieces of documents and media, and also allowing users to build and share new objects from these pieces through the Web. Aiming to better present this idea, a tool designed to extract elements such as paragraphs, tables, images and references from documents; images from videos, and graphical objects from animations, is presented. This tool also supplies automatic document construction, allowing users to immediately create and share editable documents made from other documents' pieces.

Chapter 13 - Twitter entered into the microblogging family and it is becoming a communication, information, and productivity tool. By enabling people to write messages up to just 140 characters it requires great information management and makes knowledge sharing fast and more efficient. Twitter as a learning tool can be used in different contexts. We can convert it into a tool for collaborative or individual learning. Making these assumptions the authors conducted an experiment based on microblogging with Twitter inside and outside the classroom, as a complement to Portuguese lessons. The activities were intended to help students to improve their writing skills, to develop reasoning, to improve their capacity for summarise, and to increase creativity. To achieve this students created micro stories and micro biographies, wrote Haikus, wrote a diary or poetry, published news, proverbs, thoughts, and aphorisms. These were some of the activities published by students on the class Twitter account.

Expert Commentary - Access to information technology is no longer considered a luxury but rather, a valuable resource and an important survival tool to fully participating in society (Goggin and Newell 2007; Lindsay et al. 2007; 2008a; 2009b; Mirza et al. 2006). Although access to information technology (IT) and computer ownership has quickly increased, many people still lack access to these resources (Mirza et al. 2006). Such a digital divide between the 'haves' and the 'have-nots' highlights how marginalized groups have less access to information technology. Indeed, disability is often an overlooked aspect of the digital divide and should be taken into consideration when exploring digital inequality (Adam and Kreps 2006; Arrigo 2005; Dobransky and Hargittai 2006; Mirza et al. 2006). It is vital that information technology become more accessible for people with disabilities because it can play an important role in providing information, rehabilitation and social inclusion (Arrigo 2005).

In: Internet Issues: Blogging, the Digital Divide and Digital... ISBN: 978-1-60741-824-5
Editor: Cassie M. Evans, pp. 1-25

Chapter 1

THE SCIENCE OF BLOGGING: ABOUT SCIENCE

Christine Redman
Melbourne Graduate School of Education, Science, Technology and ICT Education,
University of Melbourne, Victoria, Australia

This chapter has three main components. It provides a theoretical argument for blogging, a structure for introducing blogging into the educational landscape and a methodology to objectively code the blogging material as useful research data. The theoretical premise behind blogging in education for learning is based on the Vygotskian approach that values learning from peers in a dialectical structure. This chapter comprehensively explores the theoretical arguments underpinning the use of blogging for learning. It seeks to demonstrate how the theory informs a pedagogical structure that educators could implement when introducing blogging as and for a science learning experience. The pedagogical structure has the potential to support educators to assist the introduction as well as the progression and development of the blogging experience for learners. This focus on blogging is part of a broader perspective that Information Communication technology may be able to contribute to enhanced learning outcomes in science classrooms, though more independent engagement and the use of critical thinking skills. By utilising a socio-cultural theory of learning for the data analysis, the blogging material can be viewed as revealing the ontology of the writers. The chapter examines the contribution of blogging as a function of new technologies that may assist the enrichment of learning outcomes in science education in ways that better inform classroom practices in practical ways.

DEFINITIONS AND UNDERLYING PREMISE

The vision and goals for science education are inspiring, perhaps awe-inspiring, equally remarkable are the expectations that are being placed on the new technologies,

> In order to engage students in new and emerging science we need modern facilities and the latest equipment, technology and materials. The best learning environments will offer

> opportunities to involve students in problem solving, critical thinking and decision-making and enable them to see the relevance of science and mathematics to their world. .... Technology has the potential to increase access to education and improve teacher professional learning and student learning. .... maximise the use of information and communication technology (ICT) to create more engaging and relevant learning environments .... innovative use of ICT will extend learning experiences for students beyond the physical boundaries of the classroom .... where scientists, mathematicians and engineers are at work solving real-life problems. (Department of Education and Early Childhood Development (Vic) Policy and Research Division, 2009, p 14)

The current social constructivist approach to learning demonstrates that there is an active valuing of the perspectives of others and their co-construction, sharing and refinement of a community's ideas. It appears that aspects of ICT are anticipated to be able assist this connection between both the distant and the local community's practices. Wenger (1999) developed the powerfully clarifying and influential term a 'Community of Practice' which has highlighted the potential richness of adopting this social constructivist stance to learning.

A community of practice can be seen to develop around a common and shared point of interest and should then generate the development of a common and shared language. This concept of a community of practice has helped to support educator's visions and to consider restructuring their teaching to create small flexible working groups that have a specific and shared learning focus. It has also allowed an opportunity to support the merging of individual's interests within a mutually focused community.

The social constructivist approach could be considered to be a community constructivist approach. Blogging may help to provide an empowering vehicle to harness like-minded community in interesting ways by enabling the authors to articulate and share their understandings. The authors in a blog can also take the time required to actively craft the language they sense they need to share effectively within their community. 'All language, whether written down or spoken carries evaluative overtones' (Wolfe and Alexander, 2008).

In this chapter blogging is presented and considered to be able to provide more than an opportunity to write down ones thoughts, but also offers the thesis that it is in this dynamic process of writing and communicating with others that it seems we can help to support a more useful and purposefully focused reflection on the prior knowledge and past experiences of individuals and their community (Redman and Broom, 2009).

Overview of the Use of Blogging in a Science Education Program

What is offered here is a description of a way that the learning was staged and sequenced for science education in a higher education setting. It was designed to use a pedagogical informed approach that sought to more effectively include the use of technology tools for science education in a university level subject.

Before reaching the blogging experience I had sought to craft the learning moments that might provide individuals with both the time and opportunity to connect to the topic and also to reflect on their relevant and related experiences. These stages of engagement that occurred before the blogging commenced were considered to be critical in connecting the learner to the learning focus, while beginning to empower them as agents to be active learners.

The group were pre-service teachers, in their third year of a four year degree course to qualify them to be primary teachers. In the higher education setting there had been encouragement to use more digital technologies. A preliminary report produced in 2006 (Kennedy et al) and conducted by research teams from the Biomedical Multimedia unit and the Centre for the Study of Higher Education at The University of Melbourne highlighted the extent first year students used digital technologies. It concluded that although these students were regular users they had not yet developed the necessary practices that would assist them to effective learners.

A Curtin University research project conducted in Western Australia, in 2007, (Oliver and Goerke) reported that 'few students (7.3%) were frequent users of blogs' (2) and went on further and recommended that 'university teacher(s) ... challenge them (students) to go beyond their use purely for social ends, and use them to be participative constructors of knowledge in engaging learning experiences.'

The pedagogical strategies adopted for my teaching program were developed in response to these two studies, along with personal insights gained from other ICT related studies conducted, into effective approaches for using Information Communication Technology (ICT) to enhance teaching and learning in science education (Redman and Fawns, 2004; Redman, 2006; Redman 2008a)

The University of Melbourne had undertaken to understand what it meant to be a Digital Native in 2006. The survey about which ICT tools 2000 first-year students had used in the previous 12 months found that students had accessed the;

- mobile phone (96%)
- desktop computer (90%)
- digital camera (76%)
- MP3 player (69%)
- laptop computer (63%)

The most commonly used purposes were for;

- sending or receiving email (94%)
- creating documents (88%)
- playing digital music files (84%)
- searching for information (83%)
- communicating via instant messaging (80%).

Two of the technology tools heavily used by these students were the social communication tools of emailing and short messaging service (SMS). These both support instant sharing of ideas while still allowing opportunity for the careful crafting of a written response.

In planning my approach to this learning journey that would include blogging, I thought about the need for students to first create a strong desire to become active participants in a learning journey.

The first step involved all the students writing down their response to three questions. One question to be answered was 'what caused the moon to appear to change shape'?

The written words were then requested to be followed by a drawing which was to elucidate their written word.

In these few moments, some students discovered how inadequate their language was in serving to help them provide an explanation. Then in the subsequent moments many students also discovered that either their drawing revealed that their previous written explanations would not work, or those understandings that they had attempted to show in their drawing pointed to the fact that they were uncertain about how the science phenomena worked. This is not unusual for young children and adults, despite our fascination with the moon, it is often poorly understood, even after what appears to be good teaching (Redman, 2001; Trundle et al 2002; Mulholland and Ginns, 2008).

After the drawing experience, the students were asked to pair up with another student and when they next saw the moon, to either send an SMS message from their mobile phone, or from their computer, via email, to their partner. Most students completed this over two weeks, with their mobile phone and been able to instantly share their observations.

After two weeks had passed, the pairs of students were asked to find another pair, and informed that this group of four now would become a blogging group. This meant that the students now knew of their own individual strengths and limitations. But that they also had some stories to share from their shared experiences with a partner, and a new audience with whom they could begin their blogging conversations about their experiences. So the blog commenced with a group that had had some experience finding and watching and sharing their observations about the moon.

The research study referred to this chapter examines the development of pre-service teacher's understandings about the phases of the moon (Redman, 2008c). In that study blogging was being used to support participation in conversations around a topic that was likely to be perceived by the students to suit an out-hours approach, and could be participated in when the students were checking their email. The natural propensity of students in higher education to communicate with each other digitally was one element that initially informed the structure of the approach (Lazzari, 2007).

In the research I conducted into the contribution of blogging as a tool to enhance student engagement and learning in science education there were interesting, and unexpected, comments made by students when they were interviewed about their involvement later. Blogging was frequently perceived by learners as being associated with both contributing to their improved understandings and also recognised as having the benefit of empowering them, as blogging was perceived to be a learning style that was helpful in both sustaining and directing their focus.

This was evident when one student stated that the 'blogging experience has unquestionably refined my understandings of the moon and its relationship with the sun and the earth'. Another student reflected that although disengaged initially the process of blogging had drawn her into the task as she had come to acknowledge that she had 'discovered that this ongoing and hands-on style of learning was really well suited to me as a learner'.

The participants in the study being reported here were involved in the use of blogging to make sense of the lunar phases. The structure to the approach was scaffolded carefully to ensure that the real world had taken precedence and only after these personal experiences had been had located in the private inner space of their thinking, did the structure of the blogging attempt to provide opportunities for conversation with another, before creating options for small group discussions on the blog site. The process was designed to value both the personal experiences and the aesthetics and to support scientific inquiry by providing the time and the capacity to engage with questions of pertinence to them, as the learner.

MAIN PRINCIPLES BEHIND THE IMPETUS TO USE BLOGGING IN SCIENCE LEARNING

What is blogging? In essence a blog is rather like a form of diary writing. It is similar to a diary because the blog is usual the ideas and thoughts of a single author. Unlike a diary, a blog is generally being written to be shared. The writing of a blog provides opportunities to record one's thoughts, and inner speech, but with the objective of one's thoughts being read by others. A distinguishing feature of a blog is that there is provision for the reader to respond to the comments. So the blog can be a form of two-way communication, unlike a diary, where thoughts are not presented to others, with a blog there is perhaps an ever present expectation that there will be at least a single response or even multiple comments.

Blogging is generally classified as Web 2 technology. Web 2 technologies are generally agreed to effectively create and support information sharing and collaborations over the web. In the process of working on a blog, digital interactions and communication can take place, and this involves the two overt communication processes of writing and reading. The author writes and then someone reads the comments or ideas and they have the option of responding and offering or stating their own ideas or opinions. This element of the uploaded written response is an aspect of blogging that may promote participation and this is a distinctive feature of the blogging process.

The blogger, as the writer, understands that they are writing for an audience. Indeed, the audience can be anyone in the world, or locally known people who are part of an identifiable community. A blog may have the casual viewer who visits and comments once or not all, or may consist of regular viewers who may recurrently visit and comment on the blog. One author may have many readers and many or few interactions.

Blogging, at first glance, appears to provide simply an opportunity for participants to mediate their ideas potentially to, and with, others. But blogging may offer many learning opportunities when in the hands of the educator who is alert to how the processes of thinking and articulating are interacting and related to each other.

The action of thinking is mysterious. When we are thinking we do not need to do anything, not stop, stare, stand or sit. We can be multi-tasking and yet still be thinking, albeit about one thing at a time (Wallis, 2007). We can eat and think, drink and think, read and think and talk and think. Yet sometimes we say, 'I did not stop to think about it'. We can not see if someone is thinking or not thinking, for the process of thinking is a private place. We sometimes can 'think out loud'. This more overtly refers to the actions of talking and thinking. Thinking and language are intertwined and can be considered as inseparable.

Sometimes we may describe our thinking process as having involved us in 'thinking more deeply'. We may notice that we have had to slow down, stop, stare, tilt or turn our head, or rest our heads on our hands. Words, phrases, ideas, and sometimes images, (Arnheim, 2004; Vincent, 2007) may have taken form and can be rewound and run again, and again.

However, when we are thinking and alert to the imperative that we will be required to communicate our ideas to others we may use, and choose, our words more carefully. Sometimes we maybe actively rehearsing the words and phrases we need. The rehearsal may even be heard as a voice in our heads and the pace and intonation that will be used may have been moderated before the spoken words were uttered.

We think. We speak. We use thinking and language in our socio-cultural settings and make and take our meaning from our interactions in the world and in our community. Our words convey our intent and we convey the affordances of our participation in our world. Words serve as our tool and help us to make sense of different events and expectations. Some simple tasks may require little explanation and few words. But the

> more complex the action demanded by the situation and the less direct its solution, the greater the importance played by speech in the operation as a whole. Sometimes speech becomes of such vital importance that without it the child proves to be positively unable to accomplish the given task. (Vygotsky & Luria, 1994, 109)

In the act of blogging, as our words move into places and spaces before others in these interactions outside our private inner world, we may be called to account, to debate and define what we mean. In this process we refine our intent and meanings. Questions may be asked of us. Our words are imbued with meaning and often synonymous with our thinking, language and experiences. We have the option to either respond spontaneously or we may be required to provide a more moderated and considered answer and so this may take us more time.

The answers and responses we give may be well rehearsed and reflected upon or instant. Our response becomes and forms a dialogue as we begin to more overtly communicate our thoughts to others. Our responses to others may or may not be considered, accepted, rejected and or generate further conversation. The words and phrases we use are typically selected carefully so as to most appropriately and accurately describe our thoughts and ideas. Once this requirement or option to share becomes evident the private thoughts may be actively groomed for the public space. This shift to the public space from the private space can be perceived to be overlaid with varying degrees of accountability to others, to policy and practices.

Sometimes it may be that our spoken words may or may not align with the overtly accepted and publically stated rhetoric. In a self- selected community of bloggers, similar values and opinions are not necessarily shared. In a broader community of bloggers a wider variation in values may exist. A blogging space can provide opportunities to engage with and explore the ideas of others. The accepted socio-cultural values and views may come to be challenged, queried or just be actively sought. When our thoughts are voiced and we have articulated our ideas for others we have also moved from our inner and private safe space to a public, and potentially, vulnerable space. This action of expressing one's self publically begins the process we use to enable the mediating, and modifying, of our values and opinions and will serve to locate and identify us in a social-cultural setting (Daniels, 2001).

Experience the Requirement to Explain to Others

Try this yourself. Monitor yourself as you get prepared to both think and articulate your ideas about a science phenomenon. In the experience you are about to have you will have the opportunity to explore the process as you undertake and perhaps begin to make better sense of the complexity of this process. Can you identify, realise and recognised any shifts that you make, and have to make, as you prepare to move from your private inner world of pondering and wondering to a public space.

Be poised and primed now to examine and consider which processes you utilise and are involved in as you seek to explain what the following science conceptually laden word means to you? Ponder and consider how do you ponder the meaning? How long do you need/want to take to do this? Do you rehearse the words and phrases and where do you think these wonderings and ponderings come from? Do you begin a search for relevant examples? Do you think about and anticipate the potential effectiveness of your examples? Perhaps you selected an imaginary audience, a classroom of students you teach? Perhaps you imagined yourself explaining your ideas to a particular person, and then found that you need to re-shape and modify your description as you prepared to explain to them. The word you need to ponder for its implications and to be explained is the science concept of 'gravity'? If possible try and speak to a person or a class of students. Monitor and track your words, processes, gestures and how you felt as you gauge the responses of the listener(s).

Can you dissect and analyse the process that you undertook. Reflect on the processes that you sought to use to explain gravity. How much did assess the quality of your words, phrases and examples as you sought to be providing illuminating sentences. How do you evaluate the quality and suitability of your descriptions and examples. In those initial moments that you were responding to the question, in that instant, in those microseconds, as the words, phrases and images started to flow … you were thinking and using language… together.

At the instant and moment that you were moving from your private inner world, consider the extent to which you needed to become vigilant, monitoring the effectiveness of your efforts. Are you aware that you were perhaps also assessing the accuracy and clarity of your understandings and vigorously evaluating the extent of your knowledge? Did you draw on familiar phrases and words? Where did those come from and were they 'ready ay hand' and adequate and likely to help in the elucidation of the listener?

Did you have the option to really speak to someone? If so were the examples and language you used adequate, meaningful and helpful to the listener? When you began to speak out loud what did you say, was it the same as your rehearsal? Can you replay your comments and can you listen to each word? How did you begin 'Gravity is …? What follows may have been a mixture of words, of familiar turns of phrase, and words as well as paraverbals. So you may have started strongly, using words and phrases like forces, push and pull and then used paraverbals and proceeded with 'Um, er … um , its like, um, um, don't know …'. Only in the enacted conversations with others would you have gained a sense of the effectiveness of your explanations. It is in the conversations, in the act of sharing, in the dialogue, in the responses from others, that we can determine the effectiveness of our explanations.

We judge, we consider, how good is this, how accurate and clear are my ideas? There can also be a moral sense that arises as we take issue with, and we may have concerns about how correctly I might be covering the content. The moral factor is more likely to be present if you are the teacher and you are the person who is responsible for furthering, refining and improving understandings? So we wonder and evaluate how comprehensive am I making it or should I add more information or less? Am I understandable to the listener? We move from the private monologue in our head to a pretend dialogue with another and we have the capacity to rehearse this dialogue. We can move from our private space to a rehearsal space, then to a public space, from the individual world to a sense of the accountability of the world of the collective shared communal space. Our final spoken dialogue can be will nuanced with words, and our gestures and feelings.

The potential contribution that blogging can make lies in the use of this 'thoughtful' space; the clarifying and refining moments of the cognitive space and the accompanying language. In the move from thinking about ones ideas in one's own private space and then sharing these ideas in the public space, the thinker begins to explores and appraise their science conceptual understandings and the extent and refinement of their knowledge. The move from pre-cognitive to rehearsal space involves learners in a shift from the unreflected space to the more reflexive space.

Blogging about Science Ideas

When science educators are planning for science learning they are anticipating that the learners will become alert to the extent of their own current understandings (Driver et, 1994) and how well their existing understandings are serving them. These science understandings will often not have been explored for the extent of their usefulness, completeness, connectedness and accuracy. Science ideas have frequently not been well explored or defined as they can make up the background noise to our daily lives. People may only become aware that they are unclear about their understandings when the learners are encouraged to take the time to engage with their ideas.

So as the learner is supported to actively try to frame and explain their understandings to others, perhaps creating a shared definition, in a blog, they may become aware of the extent of their understandings.. It is in that requirement to articulate one's ideas that is perhaps the critical move for the learner, in this move towards taking ownership and responsibility for what and how well one knows something. It is the moment that the learner should be able to become alert to the boundaries and weaknesses in their knowledge, and that they can begin to take some ownership and purposeful responsibility for this vulnerability and creates a 'need to know'.

When we consider the process of blogging and align it with the processes we have just minutely explored, of thinking and language, we begin to recognise the elements that blogging may be able to offer to assist the learner in science. Before you speak you begin to assess your knowledge, and the clarity of your conceptual awareness of the science phenomenon. When you prepare to write in a blog you have to use the same processes.

In the process of blogging, in science, first the learner finds the familiar words and phrases that are related and maybe identify the relevant experiences that will explicate where these science ideas are present. Wether you are planning to speak and/or when you write you plan to the same way. You have the option to rehearse the spoken word, in the act of blogging you have the same options, and this process creates a slowing down, and more reflective place.

Think about your description and explanation of Gravity. Gravity may have been described as a dynamic force that impacts on everything you do. You may have explained that every step you take involves you in the act of falling, in response to the force of gravity, until you are able to balance yourself by placing your leg in front of you. The effect of gravity on objects is tacit and we do not need to worry or focus on it and we use the affordances (Gibson, 1997) of gravity in an unthinking way. We need to be careful to assist young learners to be able to distinguish that the invisible force is acting on objects all the time. They need to be considering the idea that this indiscernible force is constantly pulling everything towards the core of the earth. So at the moment that your thoughts and words flowed, you reflected on and again began to review, evaluate and analyse their effectiveness for communicating your ideas about gravity, and what is critical to know and understand.

The significant contribution of blogging may lie right here, for in the processes of thinking and language, in this action of writing, and in the facing of an audience, it creates a reason to explore one's ideas and probe the veracity of our understanding. Unlike the spoken word, the written word serves to slow the cognitive processes down as the time taken to write is determined by the author. They can release their writing to the outside world in their own time. The respondents can also take their time too as the responses do not need to be immediate. The writing process will support and allow the authors to ponder longer and to edit, alter and refine their written statements. The process of blogging may assist educators to slow down and deepen the thinking process as the author ponders the extent of their understandings and the connectedness of their understandings with other related events and experiences.

Blogging is more than a written artifact. It can be a reflective piece that puts ideas into the public communal space for sharing, discussion, consideration and even debate and refinement. It is an artifact written to be shared. The very act of sharing requires, perhaps even demands that the author is both clear and perhaps can be held accountable to the content. Blogging exploits the joint actions of thought and language, and the presence of an audience adds to the blogging process yet another layer of meaningfulness. The utility of Information Communication Technologies (ICT) can support and even promote the freedom and protection of multiple edits and modifications. While the author is on-line, they can research the topic and make connections or refine their understandings.

A Thesis for Blogging

Blogging can now be understood for its relationship to communication in and with a community of others. Blogging moves the individual from the world of inner private speech to a dialogical communal space with others. In the dialogic space that blogging creates we also have begun to generate a social space where science learners are often required to consider that natural everyday events may be understood in different ways (Hennessy, 1993). When learners are active participants in blogging and in the process of thinking about the extent and quality of their science ideas it seems they are being given a chance to become engaged in considering science concepts and phenomena in their personal everyday lives.

These blogging experiences become social and personal experiences that are connected to and situated in their past experiences.

Blogging promotes dialogue and exploration for learners of their own science ideas and these written conversations are distinctly different to the everyday experiences and the actions involved in shared spoken language and thinking. The written activity is being considered to be a dialogue with others, and so becomes a two-way conversation. The conversations in a blog are a place where meanings can be created, shared and refined, like a meeting with others, where understandings are being actively shared and refined (Redman, 2008a).

This idea seems so simple yet can start to be perceived as being at the centre of what may be essential in the processes involved in effective personalised learning. Blogging conversations are different as they can be likened to a slowing down of the conversation and which is one that can be crafted and refined before it is published. In a blog it is now clearly evident that possibilities exist for more preliminary thinking and refinement of the ideas that can occur before the entries are shared.

The learner has had an opportunity to take as much time as is needed to explore and participate in a deeper reflection on their science ideas. Importantly, this can occur prior to the publication events and the sharing of their understandings in a public space. In classrooms the learning environment is premised around the learners being involved in thinking. Learners are anticipated to be thinking about the designated classroom focus, and sometimes this is synonymous with the actions of listening and sitting still.

Classrooms are learning environments where thinking takes place, but thinking is invisible, so it is often not possible to 'see' if it is happening or occurring in an effective manner. The provision of time and opportunities to engage in exploratory thinking need to be catered for in the preliminary planning for teaching, and in the designated time set aside to provide for personal reflective thinking and the exploration of existing ideas. Thinking and language are part of learning and these processes can be part of classroom activities that need be seen as neither lonely nor silent.

The idea of silence being associated with thinking, points to a need to have an individual's thinking usefully illuminated by the words that might be going along with the thinking and that may indicate the person's current ideas to themselves and to others, including the teacher.

> Of all the tools for cultural and pedagogical intervention in human development and learning, talk is the most pervasive in its use and powerful in its possibilities. Talk vitally mediates the cognitive and cultural spaces between adult and child, between teacher and learner, between society and the individual, between what the child knows and understands and what he or she has yet to know and understand. Language not only manifests thinking but also structures it, and speech shapes the higher mental processes necessary for so much learning which takes place, or ought to take place, in school (Alexander, 2005).

Words and language are important to indicate which science ideas are present and how useful they are in various ways to the individual. While it is not yet clear how, or if, this is the distinctive element that is contributing to developing understandings for learners, we do generally accept the position that Wittgenstein (1958) offered us; that it is our practical experiences that give meaning to our words. Wittgenstein further suggested that it is easy in dialogue with others to misinterpret the content and intent, 'if language is to be a means of communication there must be agreement not only in definitions … but also in judgments (1958, #241). Words carry meanings and the meanings are located in the words and they can not be separated.

This chapter is not an attempt to model what thinking is, but seeks rather to assist a deeper and useful understanding of the relation of thinking and language, as inseparable, one and the same action or event.

> One way of locating thought is to think of it as a process that accompanies the uses of language and gives expressions meaning. There must, accordingly to this view, be two processes occurring when someone uses signs meaningfully: the production of the sequence of signs and the production of the sequence of thoughts that accompanies them. (Harré, 2005, p 169)

It is essential then that thought and language are harnessed and that the effectiveness of the current language that is evaluated and being utilised is considered for how helpful it is in describing the current ideas and concepts. These events in the learner's journey are important and need to have been taking place, and, before learners are encountering the thoughts and experiences of others. Once the existing ideas have been explored and deeply considered, then there can be further ideas that can be developed and it becomes possible for the learner to further explore the connectedness of one's present ideas to the ideas of others. Then eventually this should be part of an evaluation of the usefulness, accuracy and completeness of one's thoughts and language.

The complex relationship between teaching and learning may not yet be completely understood but there is an awareness that socio-cultural practices are at the critical intersection of the vortex between teaching and learning. There is the person and all that they have come to know personally, from their interactions and experiences in the world with others. Their world is known subjectively and is characterised by their experiences. Yet, their experiences are located in the social and cultural experiences of their past lived world and thus connected to the broader social world and its values.

The Inter-Relationship of the Lived World and Attitudes

The lived experiences we have from engaging in the world, and how we have experienced it, mean we have to know the world in various ways, and we have come to know it, will be different to other's experiences. So if we have never seen the sea or a rainforest then we do have the ability to make sense of it in many different ways. We may draw on different forms of media and partake of conversations, and these will better inform us of the sea or the rainforest. The experiences, one has lived, may also be associated with feelings that become associated and related to the past experiences. So if the experiences of the sea we have had have always been at the height of a hot summer, then the sea in winter could present a dramatic and even alarming contrast.

Experiences may become linked together through historical, social, cultural and emotional events. They may also be linked, inextricably, to other experiences and so become associated with other incidents and events, or not connected, where connections may be useful and informing. These associations, or not, between experiences may limit an individual's immediate ability to make connections with other related events and also may support unexpected, and unhelpful, connections.

> To be acquainted is to know what a thing is *like* in some particular. If one is acquainted with the smell of a flower it means that the smell is not just a smell, but reminds one of some other experienced thing which stands in continuity with the smell. There is thus supplied, a condition of control of or purchase upon what is present, the possibility of translating it into terms of some other trait now sensibly present (Dewey, 1906, p 296).

The personal experiences of an individual may be broadly similar to that of others yet also so broad as to be of little use when 25 students are placed together in the same classroom. The individuals who arrive on the classroom door step might all know what an apple is, but yet may have never experience eating it dried, as a juice or cooked. The extent of knowledge about the apple will vary; ideas about where it grows in the world, where it first grew, how many thousands of varieties exist and what it contributes to a healthy body may not be understandings commonly available and ready to share. Broad and varied understandings will be available to most people about apples, yet they will not necessarily have a detailed knowledge about apples, but may bring a diverse range of understandings. In the discussions that follow different ideas, based on different experiences can progressively emerge.

Once time has been provided to collaboratively explore the item called an apple then there would be a pursuing and compiling of a range of more diverse ideas, which may include historical, geographical, botanical and dietary information, and a curiosity and interest may begin to form. In the pondering about what is known and then the wondering questions may begin to follow and form. How much vitamin C is in an apple? Does the amount differ between varieties? Should apples be stored in a fridge or in a fruit bowl on a bench? Should the apple peel be eaten? When the wondering starts and the questions form and follow, it then may become more interesting and valuable to an individual to know that an apple will ripen up to ten times faster in the fruit bowl at room temperature and will stay fresh and crisp for a month in a fridge. It is useful to know that up to half to two-thirds of the fibre, vitamin C and antioxidants are in the peel of the apple and so it is best to be eaten. This knowledge contributes to an improved practical application and may reasonably be expected to assist the person to make informed choices that are contributing to the quality of their lives. This knowledge and understandings are identifiable as meaningfully and helpful to the individual.

It takes both an appropriate attitude and skill set to engage deeply in the development and progression of one's knowledge. The skill of thinking about one's existing knowledge and conceptual understandings can be progressively developed. It should enable the learner to become more skilled users of the tool set and to experience the joy and satisfaction of knowing more personalised, thus relevant and helpful information about the world. The thinker can become more strategic and focused on what is unknown and the extent to which the known is really understood. The attitudes of pondering and wondering about the natural world are aided by curiosity, and generate questions of greater personal interest. These types of personal interest questions are motivated and powered by the learner and are potentially more likely to be better resolved and understood when they evolve from this perspective.

Thinking and language are critical tools that underpin science learning experiences in the science educational learning communities that are created in the school setting, and develop vital life long skills. Schools strive to create these types of learning opportunities often and effectively. Teachers want all students to get a chance to contribute to the thinking and the conversations and to become active members of a discursive group Redman, 2005). Information Communication Technologies may be able to offer different ways of supporting educators to achieve this essentially simple and yet effective strategy.

A dialogic perspective and approach to science events may inform this educational position more effectively. Consider that 'a dialogic perspective argues that education more generally takes place within dialogic human relationships in which students see things from at least two perspectives at once, their own point of view and that of their teacher' (Wegerif, 2008). The learner in science will first needs to know what their own point of view is and then to understand the fine distinctions between their own and their teacher, and or others including the wider body of knowledge.

This is where interventions like blogging, have the potential to become useful tools for learners and particularly in science education. As the student finds that their voice is offered into their/a community, and is valued, and their experiences are useful and needed they may become more likely to adapt and adopt their understandings. In the act of blogging learners draw on any ideas that they may considered are related and associated, and they begin to be consider and contemplate and may be able to distinguish any differences that may exist between them and their immediate community, and that of the broader science community.

Once the learner has identified their own science understandings and has assessed their capacity to make sense of the relevant field, they are well positioned and empowered to take control of their specific identified learning needs. This can be seen to be a critical aspect of a learner's journey, and enhances and deepens their initial engagement with the prior knowledge. Thinking and language become the dialogic tools that support a learner to identify and refine the underlying learning needs that are present for them and importantly at the same time to become aware of how others views and or values may differ. In the act of blogging the learner has begun to develop and refine their ability to discriminate between the ontological differences in their existing knowledge.

The Role of School Science Experiments, Theories and Models and ICT

The school experience of learning about the ideas in science, in the form of science education has traditionally involved students working with experiments with the intent of getting the right results for their lab report. Not only were the right answers the goal of these experiences they were also already known to and by the teachers, and often the student. The imposed science activity was not connected to the student's need to know or interests. There has been a steady shift towards students learning from 'their doing' and this emphasis has had the teachers highlight and put more emphasis and focus on the science processes (Koulaidis and Ogborn, 1989). But there is still concern that the content-laden curriculum is distancing students, and that students do not perceive any relationship between school based science experiences and their everyday lives, needs or interests (Aikenhead, 2006; Lyons, 2006).

The latest recommendations have aimed to shift and balance the emphasis towards an alignment of the teaching of science to how it is now recognised that the students may learn science best (Bybee, 2006). The science ideas that are being included in the course designs are more likely to support the identified student's learning needs and, significantly, their learning interests and notably have some recognisable relevance to their everyday life (Rennie, 2006). It is known that through these social collaborations and mediations with other students that this can contribute to supporting the reconstruction of a learner's science perceptions and subsequently their science understandings (Tobin and Tibbins, 1993)

It has become potentially much easier to support collaborations and opportunities to work with others by including purposefully selected Information and Communication Technologies (ICT). ICT has been signaled as a key enabler in the supporting of both engagement and participation in more meaningful ways. In 1999 Hawke had highlighted that the progression in ICT tools had already provided an increase in the potential to more deeply engage students and support thinking that could be shared with others. Now with the arrival of Web 2 technologies educators are even better supported to enabling sharing and collaboration between learners.

The distinctive feature of the blogging approach that is being presented here is the strong emphasis on its potential to support a more realistic and genuinely social collaborative structure and that it can promote continued and regular opportunities for reflection (Schön, 1983) with others, who are participating in conversations that arise from a mutual place of interest. This socio-cultural pedagogical approach (Chaiklin & Lave, 1993) has been identified by bloggers in my research studies. They have noted that it has been valuable as a positive and effective contributor to their learning, 'as listening to other people form and refine their understandings filled lots of gaps in my own understandings'.

Science educators have recognised and strived to address and provide a range of diverse experiences that students have had and also that can be offered to learners in science. Classroom practices have drawn on well developed theories and practices for using investigations, models, and increased their use of ICT affordances including, in particular effective use of visualisations (Gilbert et al, 2008). Blogging is one ICT tool that may effectively harness the socio-cultural approach that has sought to provide opportunities for more prolonged interactive thinking and may create a deeper and shared engagement with the science experiences and ideas that exist differently between the communities of the learner. The same student above also explained that,

> it was through listening and participating in the blog face to face conversations that my moon understandings grew, and became more refined. I began to formulate more wonderings and ponderings on my own, as well as being inspired by other questions that the people in my blogging group had pondered'.

Blogging may offer support and a potential to better capitalise on the social shared nature of learning through its structure and can provide facilities and the ability to access a range of multiliteracies (Kress and van Leeuwen, 2001; Redman, 2009). The different forms and mixture of literacies provide access to information through sound, movement and high quality scientific graphic images and may sustain the more reiterative, personalised and stable discursive environment (Daniels, 1996) conducive to learning.

The creation of a discursive environment for science learning is based on a valuing of Vygotsky's (1987) concept that views that are communicated as the dynamic interaction between thought and words. This form of two-way communication connects the ideas of others and supports a more metacognitive and personalised learning environment. ICT, in this case, blogging, may offer a way to more easily and effectively maintain a sustained, flexible and reflexive learning environment that puts the learner's needs and interests at the centre of the focus, while drawing in and on the wider body of knowledge.

This aspect of a sustained component was identified by some of my students as being significant to them as even though concepts had been encountered throughout their schooling they had been unable to retain it. A student explained this experience as, 'We've done the lunar phases, before but a month after I had trouble explaining them. Wasn't the case this time I think because we spent so much time using different methods of looking at the moon'. So in the reiterative, multimodal approach provided in the act of blogging this student experienced success.

When a person commences to blog, they are before a computer and also online and so able to access the web and refine both their understanding and their language. The richly resourced internet can provide the science learner with access to a variety of resources that supply various ways to explore and consider their old and their new knowledge. Sometimes these diverse experiences are usefully termed multimodal. This is a helpful way to view and consider this approach. Multimodal refers to an experience which 'moves beyond language to examine how a variety of resources including colour, image, sound, spatial and kinesthetic resources are orchestrated to make meaning' (Jewitt et al, p 304, 2007).

Students identified the strength of the obvious, and sometimes latent, contributions and choices present in different media. One student stated that their learning had been enhanced, and they isolated the multimodal approach as a contributor 'this learning journey has been highly beneficial in that I was able to motivate myself to absorb new information through direct, first hand experiences and in ways that suited my own learning style – that is, visually'. Other students also signaled that while they were overtly blogging they were also accessing and using different media for example one said, 'Yes, I feel it is particularly due to playing around with different websites and models'.

In the experience of blogging the learners can enact themselves and the story of themselves, through their use of technology and their discursive agency. Through their interaction and engagement with multimodal materials that can vary from science artifacts found in the science classroom, to events and resources they find elsewhere and they may locate ICT resources that maybe combinations of textual, visual, movement and or audio.

Student engagement with these different modes includes their use of literacy skills that includes reading, speaking, listening and writing. The science learners are always located in a context, working within and in relation to the local cultural setting, and blogging and this enables a focus on not just the knowing, but also the understanding of the relationships and connections between the parts that combine to makeup the greater whole. In this active sharing and construction of a joint reality with others, the learner may begin to include a more meaningful perspective into their understanding and language. Here these learners and become able to demonstrate, present and detail their sense of conviction about the aspects of the learning journey that were successful as, 'I do believe my understandings are more stable talking with friends multiple times and modelling contributed to this I think'.

What has been identified here is that this combination of time to converse and participate in the modeling was being noted as a contributing factor to successful learning. This combination of time, as a sustained focus, and the reiterative approach blogging can support, combined with the social shared learned space was described by a learner succinctly as 'It was long term learning which we constantly revisited, it was very supportive'.

In the blogging experience the learners, may not follow the school science focus and may indeed not continue to be attending to, or limited to, just a knowledge gathering, but may be found striving towards the exploration of deeper understandings. In this study the place and use of other skills were referenced as being drawn upon and included, and the need to be exploring understandings through the modeling of concepts became highlighted. Blogging seems to support learners to more genuinely utilise and trial their ideas and rethink the appropriateness of the learning focus. The ways to learn and the activities that support learning need to adjusted to suit the learning needs and the learners need to recognise the different ways that they may get a better sense and understanding of the key ideas, 'Yes, I think modelling helps me know more about lunar phases'.

Learners Attitudes

The science process skills were being used as students observed and recorded and compared their experiences. The science processes were able to be used in blogging in ways that provided an immediate sense of being accountable to others. Attitudes to learning were also impacted upon as the bloggers worked through their understandings with others and they developed more positive attitudes as they were rewarded by the learning scenario, one stated, 'Quite confident, I'm still remembering it now, so that's good for me'.

There is a range of attitudes that can be more positively facilitated and might be recognised as significant factors that are contributing to the important group of successful lifestyle attitudes that support effective participation in a community. Blogging may provide more time for learners to refine their existing attitudes and to develop perseverance and responsibility to others.

The role of the learner may be more easily shifted from 'what does the teacher think I need to learn from the teacher/course or subject to be successful' to 'what do I need to know to help me better understand this, and what are others interested in that I might not have thought about'? The understandings maybe modified from being less about what I need to know and report on, to what do I need to understand (Bruner, 1996). In this process the tone of the attitudes, affect or motivation can be progressively changed and maintained (Ainley, 2005) and will continue to effectively contribute to the learner's participation and interest levels.

The consideration of the affective domain should be and can be required to be factored into any initiative that seeks to enhance learning outcomes for students (Redman and Coyle, 2006). The affective domain involves the values, attitudes and motivation of the learner. A focus on the affect might contribute to the support that the innovator needs to consider when trying to anticipate the impact of any interventional tool on the 'feltness' (Geertz, 1986) of the learner.

Positive attitudes motivate the learner to participate more deeply and affirm the potential effectiveness of the learning experiences and fosters their confidence to participate in and learn with and from others. These blogging students felt positive and commented that 'Yes, absolutely … remember in pictures, so doing the modelling and being 'in' the visual was really effective in helping me understand' and 'I have a much clearer and firmer understandings of the moon's phases. Absolutely. Actually doing it cemented the key concepts in my head' and,'I felt my understanding was much more stable, reviewing it so much later in the year'

In school science the constant criticism has been that learners perceive the relationship between the real world of science and school science as weak. Learners of science may have come to assume that scientific truths are waiting to be discovered and that the relationship that exists between investigations and theories is just time and opportunity. Science does not appear to relate to the learner and their lives (Rennie, 2006) and learners do not appear to understand that science is and has been a human activity process.

Thinking about science ideas takes both time and effort. Tasks that are incorporated in science activities ideally require the learners to have been connected with and involved in the natural world. The science learners need pondering and wondering time, with further time for this thinking to be followed by questions. The learners in this study seemed to have felt empowered. Their sense of agency seemed to have been developed and perhaps as learners they now may better recognise and better know how they may proceed to deeply learn about aspects of science. Their learning outcomes had been usefully connected to the real world and seemed to be highly valued. The new understandings had been experienced in meaningful ways and had become new knowledge about the natural world, 'I can really 'see' or visualise where the sun is in relation to the moon each time I look up now.'

The questions that science learners ask need to be formed into the type of questions that can be investigated or resolved, creating investigate-able questions, and this is a skill to be developed and refined by science educators. In the real world of science the journey for a new scientific idea takes time and may face repeated challenges to its claims before it comes to be accepted. Discussions, debates and arguments will prevail during this process of challenge and are part of the rigor of the scientific world.

Once an idea has been questioned and tested and the evidence explored the new idea can be accepted as a fact or theory, and sometimes only then until a refinement or challenge to the idea is next made. This discursive process of debate and discussion is connected to the testing and clarification of results and ideas and is critical in the process, and nature of science.

The ways of the science world are often hidden from the school science classroom (Sutton, 1996). But it need not remain hidden as the quest to learn and understand science phenomena can and should become part of the learning experiences that are provided in the science classroom. The genuine interest questions can be identified, shaped and pursued. Learners need to have identified their genuine need to know and it needs to be their focus that has found its place and then will drive the structure in the shared journey of the blog. This personlises the learning and eventually the science blogger experiences their voice and learning as exploratory.

In this process of using language and thinking as a describing tool, blogging progresses to become a more tentative, questioning tool. The science learner moves from a telling and descriptive voice to a more questioning and tentative voice. The blogging tool becomes a prompter of questions and supports a more analogical and exploratory style of conversation. The blogging style conversations can become less about facts and the what, and move more towards questions and wondering, and so more about sense making and less about reporting.

The reporting style language of the science classroom has been recognised as contributing to disengagement of the learner (Lemke, 1990) and is distinctly less personal and so less engaging and motivating. Reporting is also not a conversation or dialogue between and with interested others. It is also not a sharing and exploration of one's own ideas and certainly is not as a result of an interpretative journey where sense-making has been the primary objective. But this reporting style of science teaching has long since prevailed in the science education classroom.

Blogging in science creates a more personalised, responsive and purposeful structure in which conversations around meaning making can occur and science ideas can be exchanged, continually modified and refined. Blogging is a private space, which is able to be revisited on the journey of learning. It may need to be proceeded by, and intermittently interrupted by active engagement with the real world.

The real world, the natural lived world, needs to have been explored and engaged with first, before the ideas of science are considered. The meaningfulness and interest arise because of the lived world and our ontological place in it. It is the reality of our daily lives that promotes our thinking and language, it is our sensory engagement that critically underpins and informs our sense making journeys.

The real world is dynamic, often colourful, smelly and has textures and temperatures that hold us in their space momentarily, before we extract our sensory selves and think on the meaning of these experiences. It is their real world experiences that inform our later meaning making processes, and are necessary to provide the background to our ideas and subsequent understandings and it is these everyday materials, explorations and language that will first appear in our social conversations (Harré, 2002) and that educators must first ensure is present prior to the science 'lesson'.

It is the real world experiences that we have lived that forms our umwelt (Uexküll, 1982), and our subjective way of knowing our world through our engagement in it, and that will inform our way of being in the world and what we anticipate we can be doing and saying. Our umwelt informs us and shapes the possibilities and opportunities we perceive are potentially available to us to choose from and these become our associated affordances (Gibson, 1977). From these affordances we can then utilise our ways of knowing to inform our way of thinking and acting, and so how we organize our thinking and acting in a science lesson will be informed by our past.

The science education investigations and models that are presented in the classroom need to be seen to follow on from our real experiences. The science questions to be asked need to connect to and directly point to, refer to and connect us with these past ways of knowing. Blogging is a tool that can help draw upon and draw in these prior experiences and the potential affordances from our umwelt in meaningful ways. So that helps to ensure that the investigations are not used for their own sake and the models we use are better connected to what they try to represent. Blogging provides time and multiple opportunities to make sense of the arbitrary model that sp often only weakly represents the science concept being explored.

The dominating narrative that the teacher desires to have as a focus around the science model or investigation may not have primacy for the learner. The learners draw first upon their own experiences, and awareness of their own perceptions and ways of doing things (Schatzki, 2002).

Staging the Learning through a Learning Cycle

The learning cycle needs to be planned. The plan needs to involve the learner and educator taking account of the learners needs. The final structure and approach should make evident that the learner voice has been involved in development of the learning plan.

The first stage of the learning cycle is very important and significant in the long term outcomes and warrants an investment of time, from the educator and learner. The first stage of the learning cycle is at monologue, and is the moment that a topic, phrase or word elicits a response, with a set of ideas and related experiences. The first stage of the learning journey should support and involve the learner in actively reflecting on and making sense of their own ideas, including both their overt and tacit ideas. This reflective time needs to carefully planned and designed to provide the learner with the opportunity to think on what they know and bring to the fore as their first response to a topics science focus. What follows this is then the time and provision to consider how their existing ideas and experiences may relate to the ideas of others.

The learner needs to genuinely beware and to see and sense that their existing understandings are respected and valued in the learning environment. The learner should know that past ideas, experiences and understandings are an existing capacity and will be extended, or enrich or empowered in the subsequent experiences. So that this stage can progress in a useful manner learners also need to be aware that what will be equally valued in this learning environment are their subsequent questions, wonderings and ponderings. So the change here, in this approach, and needs to be signal to the learner is that this is both the moment and place to be proactive about one's understanding.

The distinctive change in this approach is that it is anticipated that the learner will be both ably supported by the educator and also by themselves. The learning objectives are not remote from the learners, the outcomes are both appropriate and the next natural stage of learning and so is notably not designed at a distance and then imposed on the unsuspecting learner. This scenario is respectful of the learners needs and will be notable as it is striving to be meaningful, and related to the learners learning needs. So this learning approach creates and addresses the point of need.

Researching the Data Set of Blogging

The most useful data analysis tool for the blogging data sets is Positioning Theory (Harré, 1999). It is 'an analytic tool that can be used flexibly to describe the shifting multiple relations in a community of practice (Linehan & McCarthy, 2000, p. 441).

Positioning Theory takes on the social constructionist nature of the social world and in these conversations it is stated by Harré that this is be the place to examine to determine what are the goals and visions of particular people, at a particular time in an institutional setting.

> The meanings of a person's actions are the acts they are used to perform. But those acts come into being only in so far as they are taken as such conversational partners. ... I don't and indeed can't decide what my actions mean. Only you and I can do that. The investigation of the devices by which some people can manage to get you to give my meaning to what both of us say and do is the study of power. (Harré, 1997, p. 182)

Harré's Positioning Theory can be used to analysis the conversational nature of this data set and the analysis will illuminate the relationship between the learner's perspective and the perceptions of the affordances being offered by the models and the blogging experience that may be helping to deepen the learners' experiences.

Harré's (2002) theory of the 'model' in science learning also considers the relation of the person to the model and the fuller encircling experiences of the learning moment. The visions, practices and articulated pedagogies that these pre-service teachers have used to make better sense of the lunar phases was being enacted as they utilise the ICT to more accurately, effectively and efficiently communicate their science understandings to others. In this data set we can a better gain a sense of what, and how, and if, they were actively developing a better science learning outcome by using ICT for modelling science ideas and the ICT component of blogging to discursively share their ideas.

It is through the written words that will make sense of their understandings as, 'Often some of the most revealing analytical insights come during the transcription because a profound engagement with the material is needed to produce a good transcript.' (Potter, 1996)

This ontological approach to the teaching of lunar phases needed a methodology that took a social–psychological approach to the data set and Positioning Theory supports this approach (Redman, 2007). In order to identify and describe the range of perceptions and affordances that were experienced with the different models that the learners used, the blogging data is excellent as a first place of call, but it also needs a conversation around the data set for it to offer the more insightful aspects.

Pronominal tracking will support a richer understanding of the degree of commitment and confidence that people had with their new ideas as they were being developed. Harré's Positioning Theory is a dynamic alternative to the concept of role that can be used to make more determinate the storylines, positioning and illocutionary force of the everyday conversation and blogging scripts. By tracking the changes in the pronoun grammar the researcher can better see the changes in the affordances and, importantly, relationships with the content knowledge being developed.

The methodological tool of pronoun grammar analysis helps to focus the researcher on the place of the learner in the conversation (Redman, 2008b). This study of the dialogue reveals how individuals are socially constructing a self, forming and reforming their personal identity, while interacting with others in the blog. The analysis of the blog can begin to indicate what it is that is being enabled or constrained in the learning.

One of the key indicators for changes in a social location is the choice of the pronoun. In a conversation there is a choice between using 'I' or 'you'. The use of 'I' refers to Self 1. Self 1 is the person that oneself sees and refers to which is perceived as distinct from anyone else. Harré refers to this as 'the reification of the unique personal point of view' (1999, p.81). The use of 'I' indicates that possibly the individual has moved the object/artifact under discussion into their real material world. Whereas the use of 'you' may indicate that the object is being viewed as outside the immediate, omnipresent and meaningful world of the user.

The blogging data set is an interesting set of data. It will often be written in a personal style, like a diary, and like a diary it can tend to reflect musings and thoughts. This is perhaps a key strength of blogging as a learning tool, as it allows the new understandings to be stated and then considered, rather than statements being made and learners feeling committed to them. The language is often exploratory and very personal. In this more exploratory place, with others learners, they can probe their understandings and better make sense of the learning journey, taking more responsibility for their learning.

THE PRACTICAL ASPECTS OF THE E-CONVERSATION AND ITS CONTRIBUTION TO PERSONAL IDENTITY FORMATION

The blogger has used electronic words in their communications and unlike the hand written word the electronic text can be drafted, considered and easily quickly edited and re-edited. Also, unlike the written word the electronic word can be stored and when necessary it can be quickly found and then reviewed for its ideas. This ability to locate the text and easily be able to be retrieved is important for learners in science who can review their learning journey. They can check that they have covered the items that they were first interested in and be confident that they are covering the questions they had flagged initially.

Another significant useful function of the electronic text is that it can be searched. Hand written work is too cumbersome to be searched for relevant and particular items. Electronic documents can be searched for the relevant and desired information. Even parts of words can be located and an accurate comprehensive search made of the text. In the blogging scenario this ability to be able to track and retrace the steps and stages of the learning can be significant. The learner can identify their place in the learning journey and access a potential record of their progress.

The electronic word can be interacted with, in real-time, so a response or comment can be posted and feedback gained in a moment. But slower responses may arrive, as these responses have been researched and considered in different ways, perhaps sometimes but not always while, live on the internet. The blogging response may return and have with it photos or videos attached or maybe it provides links to relevant information. The electronic word has become pervasive and is readily available and cheap and efficient to use. The numbers of members in a blog can be constrained to a few or extended to larger numbers. It provides the freedom to share rich resources to many or a selected group of significant others, and in addition that can be done anytime and anywhere.

In this writing process where the author extends their agency and makes their choices and deliberations about what they will say or ask, in this purposeful communication the author as the writer begins to negotiate themselves and their identity. This is a dialectic process in which the author works on themselves as the self-as-product of social positioning and the self-as-process and as cultural agent. The journey involves the author working on them self, as thinker and as questioner, and as learner and sometimes as teacher.

There is a conscious consideration of what others expect of me and what I would like to do or I am capable of doing. People speak from this sense of 'oughtness' (Linehan and McCarthy, 2000), and take up new understandings and practices based on the extent of their awareness that this is what 'we', in this profession or institution perceive 'we' ought to be doing and knowing and valuing. In this dynamic and yet collaborative process of sharing one's writing, individuals perhaps define their identity and also, very importantly, redefine it 'through negotiation of the meanings of our experience of membership in communities' (Linehan & McCarthy, 2000, p 438).

You and Your Student's Blogs

In the interviews held after the blogging experience had concluded, one student stated that they were conscious of my actions as a moderator, although they knew I would not comment into their blog. But another student stated that they were not conscious of my presence and when they were writing they felt they were writing just to their group.

In lectures I would make reference to interesting comments, observations and sharing that had been going on in the blogging groups. But I did this without mentioning the group, and the groups enjoyed this and would say to each other in the blog space, was that a reference to us? Sometimes, without identifying the group, I could offer thoughts for consideration and to expand their ideas, questions and focus. This too would appear in the blog, and they would ask each other prompting questions, like 'so does that mean …?'

The blogging process still needs to be refined for its use in education, and investigated, in a range of settings. Researchers need to deliberate seek to determine the extent of its contribution as a learning tool that maybe capitalising on this generation's ability to communicate their thoughts and ideas. The digital affordances of the ICT and social collaborations and real world personalised learning experiences are all aspects that warrant further time and consideration to maximize the opportunities available here.

References

Aikenhead, G. (2006). *Science education for everyday life*. New York: Teachers' College Press.

Ainley, M., Corrigan, M. & Richardson, N. (2005). Students, tasks and emotions: identifying the contribution of emotions to students' reading of popular culture and popular science texts. *Learning and Instruction*, *15*, 433-447.

Alexander, R. (2005). Culture, Dialogue and Learning: Notes on an emerging pedagogy. Keynote, Education, Culture and Cognition for growth, *International Association for Cognitive Education and Psychology* (IACEP), Durham, United Kingdom.

Arnheim, R. (2004). *Visual thinking*. Berkeley: University of California Press.

Bruner, J. (1996). *The Culture of Education*. Cambridge, MA: Harvard University Press.

Bybee, R. (2006). Enhancing science teaching and student learning: A BSCS perspective. ACER Research Conference 2006, '*Boosting science learning-What will it take?*' Canberra, Australia.

Chaiklin, S. & Lave, J. (1993). *Understanding practice: perspectives in activity and context*. Cambridge: Cambridge Press.

Dewey, J. (1906). The Experimental Theory of Knowledge. *Mind, a quarterly review of Psychology and Philosophy*, *59,* 293-307

Daniels, H. (1996). *An Introduction to Vygotsky*. London: Routledge.

Daniels, H. (2001). *Vygotsky and Pedagogy*. London: Routledge/Falmer.

Department of Education and Early Childhood Development (Vic), (2009). Policy and Research Division, *Energising Science and Mathematics Education in Victoria;* Blueprint Implementation Paper, DEECD, Melbourne. http://www.education.vic.gov.au

Driver, R., Asoko, H., Leach, J., Mortimer, E. & Scott, P. (1994). Constructing scientific knowledge in the classroom. *Educational Researcher, 23*, 5-12.

Geertz, C. (1986). Making experiences: authoring selves. In: E. M. Bruner, & V. W. Turner (Eds.), *The anthropology of experience*. 373-380. Chicago: University of Illinois.

Gibson, J. J. (1977). The theory of affordances. In R. Shaw, & Bransford, J. (Ed.), *Perceiving, Acting and Knowing*. Hillsdale, NJ: Erlbaum.

Gilbert, J., Reiner, M. & Nakhleh, M. (2008). *Visualization: Theory and Practice in Science Education*. New York: Springer.

Davies, B. & Harré, R. (1990). Positioning: The Discursive Production of Selves. *Journal for the Theory of Social Behavior*, *20(1)*, 43-63.

Harré, R. (1997). Forward to Aristotle; the case for a hybrid ontology. *Journal for the Theory of Social Behavior, 27(2/3)*, 173-191.

Harré, R. & van Langenhøve, L. (1999). *Positioning Theory: Moral Contexts of Intentional Action*. Oxford: Blackwell.

Harré, R. (2002). Material objects in a social world, *Theory, Culture and Society*, *19(5-6)*, 23-33.

Harré, R. & Tissaw, M. (2005). *Wittgenstein and Psychology: a practical guide*. United Kingdom: *Ashgate Publishing*.

Hawkey, R. (1999). Exploring and investigating the natural world. Sc1 on-line. *Primary Science Review*, *60,* 4-6.

Hennessy, S. (1993). Situated Cognition and cognitive apprenticeship: Implications for classroom learning. *Studies in Science Education, 22*, 1-41.

Jewitt, C. Moss, G. & Cardini, A. (2007). Pace, interactivity and multimodality in teachers' design of texts for interactive whiteboards in the secondary school classroom. *Learning, Media and Technology*, *32,* 3, 303-317.

Kennedy, G., Krause, K. L., Judd, T., Churchward, A. & Gray, K. (2006). *First year students' experiences with technology: Are they really digital natives?* Melbourne, Australia: Biomedical Multimedia Unit, University of Melbourne. http://www.bmu.unimelb.edu.au/ research/munatives/natives_report2006.pdf.

Koulaidis, V & Ogborn, J. (1989). Philosophy of science: an empirical study of teachers' views. *International Journal of Science Education, 11(2)*, 173-184.

Kress, G. & van Leeuwen, T. (2001). *Multimodal Discourse*. London: Edward Arnold.

Lazzari, M. (2007). Podcasting in the classroom: involving students in creating podcasted lessons. *HCI Educators conference*, Aveiro, Portugal.

Lemke, J. L. (1990). *Talking Science: Language, Learning and Values*. Norwood, NJ: Ablex.

Linehan, C. & McCarthy, J. (2000). Positioning in Practice: understanding participation in the social world. *Journal for the Theory of Social Behaviour, 30*, 435-453.

Lyons, T. (2006). The puzzle of falling enrolments in physics and chemistry courses: Putting some pieces together. *Research in Science Education, 36(3)*, 285-311.

Mulholland, J & Ginns, I. (2008). College MOON project Australia: Preservice Teachers Learning about Moon's Phases. *Journal of the Australian Science Education Research Association, 38(3)*, 385-399.

Oliver, B. & Goerke, V. (2007). Australian undergraduates' use and ownership of emerging technologies; Implications and opportunities for creating engaging learning experiences for the Net Generation. *Australasian Journal of Educational Technology, 23(2)*, 171-186

Potter, J. (1996). *Representing Reality, Discourse, Rhetoric and Social Construction*. London: Sage.

Redman, C. (2001). Moon rise, Moon set. *Investigating, 1(17)*, 22-7.

Redman, C. (2005). Planning Professional Development: meeting the needs of teacher participants. In: S. Rodrigues, (Eds.), *International Models of Teacher Professional Development in Science Education: Changes influenced by politics, pedagogy and innovation*. New York: Nova Science Publishers, 189-209.

Redman, C. & Fawns, R. (2004). Discursive positioning and effecting change in a community of practice. *Australian Association of Research in Education* (AARE), Melbourne, Australia

Redman, C. & Coyle, D. (2008). The place of the social, cultural and affective domains in supporting the uptake of new technologies. *International Federation for Information Processing, (IFIP), WG 3.5 Informatics and Elementary Education. Valuing Individual and Shared Learning*, Prague, Czech Republic

Redman, C. (2007). Towards a dialogical perspective on agency in student learning. *Australian Association of Research in Education* (AARE), Fremantle, Australia

Redman, C. (2008a). The Research Planning Meeting. In R. Harré, F. Moghaddam & N. Lee (Eds.), *Global Conflict Resolution Through Positioning Analysis*. 95-112, New York: Springer.

Redman, C. & Rodrigues, S. G. (2008). Researching the relationships in the Technologies of Self: Habitus and Capacities, *Australian Association of Research in Education*, (AARE). Brisbane, Australia.

Redman, C. (2008b). Contributions of video-podcasting and blogging to increase reflection in science education learning at tertiary level, In Proceedings of World Conference on *Education*al Multimedia, Hypermedia and Telecommunications 2008 (3908-3916). Chesapeake, VA: AACE

Redman, C. (2008c). Preparing the stage for using 'emerging technologies' in science education, World Computer Conference, 2008, (WCC), TC3, *Learning to live in the Knowledge Society Conference*, Milan, Italy.

Redman, C. (2009). Pedagogical approaches for using multimedia and multiliteracies in science education, *Redesigning Pedagogy: Designing new learning contexts for a globalizing world,* Singapore.

Redman, C. & Broom, L. (2009). Transforming reflective practices in clinical practice through discursive electronic technologies, *15th International Reflective Practice Conference*, Limerick, Ireland.

Rennie, L. J. (2006). The community's contribution to science learning: Making it count. ACER Research Conference 2006, '*Boosting science learning – What will it take?*' Canberra, Australia.

Schatzki, T. (2002). *The site of the social: A philosophical account of the constitution of social life and change*. Pennsylvania: Pennsylvania University Press.

Schön, D A. (1983). *The Reflective Practitioner: how professionals think in action.* London: Temple Smith.

Sutton, C. (1996). Beliefs about science and beliefs about language. *International Journal of Science Education, 18(1),* 1-18.

Tobin, K. & Tippins, D. (1993). Constructivism as a referent for teaching and learning. In K. Tobin (Eds.), *The practice of constructivism in science education*, 3-21. Hillsdale, NJ: Lawrence Erlbaum.

Trundle, K. C., Atwood, R. K. & Christopher, J. E. (2002). Preservice elementary teachers' conceptions of moon phases before and after instruction. *Journal of Research in Science Teaching, 39(7)*, 633-658.

Vincent, J. (2007). Literacy, learning preferences and multimedia. *Language and Education, 21*, 141-157.

Wallis, C. (2006). The multitasking generation. *TIME*. http://www.time.com/time/printout/0, 8816,1174696,00.html.

Wegerif, R. (2008). Dialogic or dialectic? The significance of ontological assumptions in research on educational dialogue. *British Educational Research Journal, 34(3)*, 347-361.

Wenger, E. (1999). *Communities of Practice: Learning, Meaning and Identity.* Cambridge: University Cambridge.

Wolfe S. & Alexander, R. (2008). Argumentation and dialogic teaching: alternative pedagogies for a changing world. *Beyond current Horizons,* http://www. Beyondcurr-enthorizons.org.uk/argumentation-and-dialogic-teaching-alternative-pedagogies-for-a-changing-world/.

Uexküll, von J. (1982). The theory of meaning. *Semiotica*, *42(1)*, 25-82.

Vygotsky, L. S. (1987). *The Collected Works of L.S. Vygotsky. Vol. 1. Problems of General Psychology.* Plenum Press: New York.

Vygotsky, L. & Luria, A. (1994). Tool and symbol in child development. In R. van der Veer & J. Valisner, (Eds.), *The Vygotsky reader.* Cambridge, MA: Blackwell.

In: Internet Issues: Blogging, the Digital Divide and Digital... ISBN: 978-1-60741-824-5
Editor: Cassie M. Evans, pp. 27-52

Chapter 2

A HYPERLINK STUDY OF GREEK POLITICAL BLOGS COMMUNICATION PATTERNS

Kostas Zafiropoulos[1] and Vasiliki Vrana[2]
[1]Department of International and European Studies, University of Macedonia, Greece
[2]Department of Business Administration, Technological Education Institute of Serres, Greece

ABSTRACT

Blogs are web pages that accept comments-posts of users in the form of a diary. They allow posting and exchange of opinions and are part of WEB 2.0. Posting of opinions in the Internet can be considered as an expression of activated public opinion in contradiction to public opinion, which is recorded through surveys and concerns the public. The chapter starts by providing a literature review on blogging and political blogging. It presents the recent advances in the hyperlinks analysis of blogs, and describes the recent hypotheses used for this kind of research. As a case study, the chapter examines the degree of interconnectivity, cohesion and polarisation of Greek political blogs that participate in the discussion and use tags about Greek Parties. Tags are labels of thematic categories of blog posts regarding their subject. In order to cope with the research question, the study records Greek blogs with tags about parties of the Greek Parliament. Blogs with tags about Greek parties ND, PASOK, KKE, SYRIZA, and LAOS are considered. This chapter analyses connectivity patterns of Greek political blogs using a two step approach. First, it uses a Social Networking Analysis of the blogs using hyperlinks. Graph theoretic properties and indexes are used to present connectivity properties of the blogs in order to explore the fashion of blogs self-organization. The second step regards the exploration and identification of focal conversational points. The basic hypothesis supported by the literature is that within polarized political systems blogs are forming clusters around central blogs, which are considered reliable or having the same affiliation. Users of the Internet who wish to be informed quickly, locate the focal points of discussion, and for economy of navigation they read only the posts on these blogs. Bloggers also locate focal point blogs and place their posts along with a link to their blog. They expect thus that the readers of focal point blogs will also visit their blogs. The research examines whether this hypothesis holds for Greek political blogs.

The quantitative analysis initially uses the theory of Social Networking. The blogs and their incoming links along with the association matrixes (0/1) of the blogs networks are recorded. Multidimensional Scaling and Hierarchical Cluster Analysis are used to locate clusters of focal point blogs. Several quantitative indexes are calculated to present the degree of interconnectivity, cohesion and polarization of blogs tagging to Greek Political Parties. Regarding the Social Network properties of the blogs, the Greek political blogosphere is connected, though not presenting high density. Regarding the formation of central blog groups, Greek political blogs conform well, though not completely, with the communication pattern supported by the literature. Study of the political and ideological affiliation of the central blog clusters, along with the study of the volume of connectivity (through hyperlinks), describes a system, which is partisan but much polarized.

Keywords: blogs, Greece, political blogs, blog authority, links distribution, skewness, connectivity, patterns, social networking, cluster analysis, Multidimensional Scaling, focal points, core groups, partisanship, polarization.

INTRODUCTION

Blogs are low-threshold tools for Internet users to personalise and actualise content and information online and to present their views to a broad audience. They help self-expression and self-empowerment, are becoming a major source of information and communication for Internet users and have the power to engage people in collaborative activity, knowledge sharing, reflection and debate (Blood, 2002; Efimova et al., 2005; Hiler, 2002; Punie & Cabrera, 2005). Blogs are not only employed in personal environments but also in organizations and enterprises (Kolbitsch & Mauer, 2006). They are often perceived as powerful because they allow millions of people to easily publish and share their ideas, and millions more to read and respond (www.technorati.com) and as a good way of circulating new ideas (Baker & Green, 2005).

Blogs are also featured in papers regarding their political activism and influence (Su et al., 2005). According to Drezner & Farrell (2004) there is evidence that blogs have affected real world events. Moreover, blogs serve as a medium and tool for political change and many times are creating the media agenda (Cornfield et al., 2005). The reference point for the political impact of blogs was the 2004 U.S. presidential campaign (Albrecht et al., 2007). Thereof blogs are used as a source of information for the purpose of political debate, to mobilize supporters and funding (Kerbel & Bloom, 2005) and to influence the news, political and policy agenda (Jackson, 2006).

Nowadays, blogging tools provide enhanced features for between-blog interactivity; thus promoting the creation of social networks among bloggers (Du & Wagner, 2006). Blogrolls, permalinks, comments and trackbacks are the technology features that facilitate interactions between bloggers and visitors, and also enhance conversations among bloggers. Williams & Jacobs (2006) recognized interactivity as the key to the success of social network systems such as blogs. However a small set of blogs, the A-list blogs, account for the majority of the traffic in the blogosphere and the others suffer from poor readership (Drezner & Farrell, 2004; Guo et al., 2008 ; Nardi et al.,2004).

Bloggers are not a homogenous group, but they are an educated and affluent one. According to Technorati.com (2008) three out of four U.S. bloggers are college graduates, and 42% have attended graduate school. They skew male, and more than half have a household income over $75,000. Bloggers are also experienced. Half of bloggers are on their second blog, and 59% have been blogging for more than two years.

The chapter aims to describe basic characteristics of political blogging in Greece using at large their quantitative properties regarding their hyperlinks

Blogs

Barger (1997) used for first time the term weblog and defined blog, short for weblog, as ''a web page where a blogger 'logs' all the other web pages he finds interesting''. Drezner & Farrell (2004, p. 5) defined weblogs as "a web page with minimal to no external editing, providing on-line commentary, periodically updated and presented in reverse chronological order, with hyperlinks to other online sources" and Kolari et al. (2006, p.92) as "websites consisting of dated entries typically listed in reverse chronological order on a single page". Gumbrecht (2004) characterized blogs as personal protected space where the author can communicate with others while retaining control.

Even though blogs can take a number of forms, they tend to have a number of characteristics in common. Blogs are being set up and controlled by the author, display blog postings in reverse-chronological order or arranged by subject. Each entry has a timestamp so that the reader knows when it was posted and to alert readers to the "currency" or "timeliness" of the log. They are updated regularly; some bloggers update several times a day, while others may update every few days or once a week with new material. Blogs usually tend to cover a very specific subject area, provide a personal viewpoint and facilitate critical feedback by giving their readers the opportunity to comment on blog postings, or to contact the author directly (Lankshear & Knobel, 2003; Pedley, 2005).

Everyone can set up a blog in less than ten minutes with a minimal cost (Drezner & Farrell, 2004). Pedley (2005, p.95) mentioned "A major attraction of weblogs is their relative ease of construction/updating and the lack of the need or any special skills in web design or of HTML coding" and Jackson (2006, p. 294) wrote that with blogs "even technophobes could get online". Rosenbloom (2004) claimed "a blogger only needs a computer, internet access and an opinion". However, it needs effort to maintain an "active" blog. Activity depends on two parameters: the blogger, who needs to update the content regularly, and the readers who need to visit and interact often with the blog. It is a common phenomenon to abandon blogs soon after their creation (Hsu & Lin, 2008).

Regarding types of blogs, Blood (2002) and also Herring et al. (2004) distinguished three types: personal journals, filters and notebooks. In personal journals, the content is internal by means that the blogger express thoughts and daily activities. The content of filters is external to the blogger. They are pointing to content that the blogger finds interesting and provide commentary on it. Finally, notebooks may contain either external or internal content but are distinguished by longer, focused essays. At another approach Krishnamurthy (2002) classified blogs into four basic types according to two dimensions: personal vs. topical, and individual vs. community.

As blogs are growing in popularity, businesses and organizations are looking for ways to exploit them. In the business world, blogs are considered as environments for knowledge sharing (Festa, 2003), a unique opportunity for informal knowledge sharing as described by Kosonen et al. (2007), a "magic" formula for corporate communication (Jüch & Stobbe, 2005), and a potential for future profit (Lu & Hsiao, 2007). Companies like Microsoft and IBM encourage employees to embrace blogs (Du & Wagner, 2006) and Disney shares knowledge through an internal team blog (Guo et al, 2008). Companies also use blogs in order to become recognized in the industry and to take direct feedback from customers, by allowing the public to make comments on blog posts (Hepburn, 2007). Southwest Airlines and Starwoods Hotels and Resorts maintain such official blogs (Dwivedi et al, 2007). Moreover, companies can use selected blogs for advertisement. Companies such as Nike and Paramount utilise blogs as a new way to reach potential customers (Hsu & Lin, 2008). Amazon.com uses a blog in order to announce product enhancements and provide support (Guo et al., 2008). Fuchs (2007) highlighted blogs' potential to advance cooperation and participation in social systems and Hoogenboom et al. (2007) their support to social learning processes. In education, blogs can help students to share their learning experiences and express their thoughts to the instructor (Maag, 2005), facilitate extended discussions beyond class meetings (Betts & Glogoff, 2004), increase students' intellectual exchange (Williams & Jacobs, 2004) and finally enhance their studies (Kim et al., 2008). Librarian blogs serve as new information channels both for the library and information science community and for the general public (Bar-Ilan, 2007).

Blogs and Journalism

Usually blogs are maintained by individuals to put forward their personal thoughts, however there are many blogs written by experts within a particular subject. A question that arises is whether blogging is journalism. Definitely, not everyone who keeps an on-line journal is a journalist. Andrews (2003) wrote ".... it is fair to say the vast majority of blogging does not qualify as journalism. If journalism is the imparting of verifiable facts to a general audience through a mass medium, then most blogs fall well short of meeting the standard". However, blogs are a form of journalism and are changing the way in which journalism is practiced today. Lasica (2002) and Pedley (2005) characterized blogs as "amateur" journalism and Pedley (2005) wrote "This can be seen as part of their strength but also as one of their main weaknesses". Wall (2004) characterized them as "informal journalism" and "black market journalism" because they are not operating within the standard set of rules of formal journalism and she highlighted the fact that blogs are attempting to bypass traditional media news channels. Commenting on this Allan (2002) mentioned that news blogs unlike mainstream media are generally not objective or detached but opinionated and personal, and Lasica (2002) mentioned "... the most vital and moral dispatches on the Web are being created by amateurs". There is an open argument between bloggers and journalists about how much faith to place on blogs (Johnson et al., 2007). Even though blog users are aware that bloggers are opinionated, they trust blogs and this gives blogs power and bring about social change (Banning & Trammell, 2006; Johnson & Kaye, 2004; Johnson et al., 2007). Especially politically interested Internet users rely more on blogs than on any other news source for news and information and judge blogs as more credible than either portal or mainstream news sites (Johnson & Kaye, 2007). Controversial opinions also exist. According to Koh et al., (2005, p. 2) "bloggers are often criticized for not being "professional journalists," with the assumption being that they have neither the training nor the judgment necessary to present news and opinions that could affect public opinion.

Journalists regard blogs as alternative sources of news and public opinion (Lasica, 2001) even thought they view blogging sceptically (Williams & Jacobs, 2004) and perhaps a threat to their profession (Thompson, 2003). Many journalists keep their own blogs and very few of them are paid by their organizations to do so. However, the majority of journalists do not keep blogs. (Andrews, 2003). Robinson (2006, p.79) claimed that journalists' blogs are the corporate answer to independent bloggers and she mentioned "They are a way for journalists to reclaim journalism – and its standards – online, even through a postmodern entity originally created to defy those traditional norms". A few newspaper blogs also exist, which are very popular. Pohlig (2003: 25) wrote on this "....[journalist blogs] are hugely popular, largely because they allow the reader to see the journalist as a human being, connecting with them without the stiff, imperial we voice that turns so many younger people off".

Political Blogging

Political blogs are a new form of political expression and participation and have the potential to shape politics and political discourse (McKenna & Pole, 2004; Wallsten, 2005). Political blogs have been described as an alternative channel for the distribution of information, as a mobilization tool due to their ability to spread news very quickly (Albrecht et al, 2007 ; Kahn & Kellner, 2004), as a credible source which provides depth and thoughtful analysis (Andrews, 2003; Johnson & Kaye, 2004; Regan, 2003), and as "gatekeepers of information and news" (Pedley, 2005 p. 97). Herring et al. (2004) claimed that political blogs possess a "social-transformative, democratizing potential" and Walker (2007) called blogs "a soapbox" and highlighted their attractiveness for voicing political messages.

Blogs were multiplied exponentially after 11/9 as people used them to express their political awareness (McKenna & Pole, 2004). There is an expanding number of situations in which blogs have exercised an important influence over how politics are practiced and policy is developed and have broken major news that sometimes contradicted the point of view of the mainstream media (Jackson 2005; Lankshear & Knobel 2003; Sroca, 2006). "The interest within the political sphere on bloggers is that they are a potential alternative to the traditional media" mentioned Pedley (2005, p.295). Moreover, blogs have the advantage of speedy publication and there are many paradigms that political blogs provide news that is not readily available elsewhere. In September 2001 bloggers created a protest against Trent Lott who supported Strom Thurmond's segregationist stance in the 1948 presidential election at the 100th birthday of Senator Strom Thurmond (Bloom 2003). Bloggers also echoed the suspicion that "President Bush was using a listening device for assistance during the presidential debate and worries over the validity of voting machines" (Su et al. 2005). In November 2005, bloggers in Germany found that the picture used for the claim of the social marketing campaign had already been used by some Nazis in 1935 (http://www.spreeblick.com/2006/01/27/you-are-deutschland-too-just-kidding/). In the US-led invasion of Iraq, blogs posted photos before other media and sometimes contradicted the point of view of mainstream media like CNN (Lankshear & Knobel, 2003). The famous nowadays blogger Iraqi Salam "gave outsiders a dose of the larger unexpurgated reality as the bombs exploded overhead in Baghdad" and Hossein Derakhshan became the first blogger to be jailed for "undermining national security through cultural activities" (Kahn, R. & Kellner., 2004, p.90).

Political parties use blogs in order to shape the political agenda, generate resources, mobilize support and reach out to their constituency (Trammell et al. 2006). According to Chadwick (2008) primary and presidential campaigns in the United States "saw the emergence of campaigning model based on online venues loosely meshed together through automated linking technologies, particularly blogs". Jackson (2006 , p.295) suggests that "during an election campaign a weblog is a mean for a party to promulgate its policies through a virtual network of political bloggers". During election campaigns parties and candidates blogs are used for three main functions: diffusing information to internal audiences; building up a volunteer base; and agenda setting of the mainstream media (Bloom & Kerbel, 2005). However, a few of the political blogs have been run by political candidates (McKenna & Pole, 2004). Albrecht (2007, p.504) claimed that blogs "offer forms of communication that allow political actors to bypass established media practices". An example of a highly successfully blog is Howard Dean's Blog for America. The blog was used to mobilize supporters and funding, even though Howard Dean's candidacy was not successful (Albrecht et al., 2007; Kerbel & Bloom, 2005; McKenna & Pole, 2004). Blogs have also raised money for candidates in elections in some congressional districts in USA. In Ohio, for example, even thought Paul Hackett lost in elections to Republican Jean Schmidt, his contests have been far more competitive than expected. In South Dakota Politics, Jon Lauck and Jason Van Beek were credited with aiding John Thune's victory over Senate Minority Leader Tom Daschle, D-S.D. Thune's campaign reportedly paid both bloggers, and Thune later hired Van Beek as a Senate aide (Glover, 2006). According to Albrecht et al. (2007) paradigms that blogs are used as campaign instruments are the 2004 presidential election, the 2005 U.K. general election, the 2005 Danish parliamentary election, the 2005 New Zealand general election, the 2005 German Bundestag election, the 2007 French election and the 2007 Australian Federal Election. At recent presidential elections in USA, Barack Obama maintained a blog at his webpage (http://www. barackobama.com).

Investigating political blogs for campaigns, Garrett (2004) (in Trammell et al. (2006)) mentioned that bloggers write posts in a personal voice, update the blogs several times a day, encourage and moderate comments, offer hyperlinks to internal and external sources and other blogs, and call the readers into action. Graf & Darr (2004) pointed out that blogs appear to play an increasingly important role as a forum of public debate, with knock-on consequences for the media and for politics. "In just a few years they have become a finger in the eye of the mainstream media and a closely watched forum of political debate. Political blogs have exposed lapses in mainstream media coverage, chastened reporters with the fear of an angry online response to sensitive stories, and at times set the media agenda. Political blogs have also been influential in raising money for political candidates and pushing select races into the national spotlight" (Graf, 2006; p.3)

Graf (2006) tried to investigate the audience of political blogs and undertook an online survey. He used a sample of 7,863 people and they mentioned that a "minority of our sample has at least seen a political blog. About 40% of everyone sampled said they had looked at a blog "that discusses politics or current events" at least once in the past month. Another 7% said they visited "several times a week." However, only 9% of our sample respondents said they visited political blogs "almost every day." The findings of the survey saw that the daily audience for political blogs is fairly small however. Bloom (2003) also commented about visitors of political blogs in USA, and mentioned that a high percentage is political reporters, politicians and policy makers: key opinion formers. Taking this in to consideration, Jackson (2006, p.296) mentioned: "This can give political bloggers a disproportionate influence, based on the type of blog visitor, and not just the number of blog visitors. Therefore, elite bloggers can act as a "focal point" encouraging influential visitors to congregate around them. To influence the news, political and policy agenda, political actors need to attract an "A" list audience to their weblog".

In Greece however, where the ratio of internet users is relatively small, there is an expanding portion of bloggers who comment regularly and have the power, to a certain degree and in certain circumstances, to trigger off political movements. Posting of opinions on the Internet can be considered as an expression of activated public opinion in contradiction to public opinion, which is recorded through surveys and concerns the wider public.

Blogs Hyperlinking

By definition, blogs link to other sources of information; usually to other blogs. The most important difference between blogs and more traditional media "is that blogs are networked phenomena that rely on hyperlinks" (Drezner & Farrell, 2004; p.5) and on this depend their success (Williams & Jacobs, 2004). Blogging tools provide appropriate features for managing blog interactivity and promoting the creation of social networks among bloggers (Du & Wagner, 2006). Sigala (2008) mentioned, "blogs create and maintain strong online communities through their social ties tools such as blogrolls, permalinks, comments and trackbacks".

A "blogroll" is a list of blogs that many bloggers maintain. The blogroll occupies a permanent position on the blog's home page and is the list of blogs that the blogger frequently reads or especially admires and thus offers links to these blogs. Blogrolls evolved early in the development of blogs and serve as a navigation tool for blog readers to find other blogs with similar interests (Marlow, 2004). Commenting on this Drezner & Farell (2004, p.7) wrote: "Blogrolls provide an excellent means of situating a blogger's interests and preferences within the blogosphere. Bloggers are likely to use their blogrolls to link other blogs that have shared interests". Albrecht et al. (2007, p. 506) referred to this form of interactivity between blogs as "connectedness of weblogs".

Interactivity can also be achieved with permalinks and comments. Comment tools are allowing readers to express their opinions on a weblog entry and permalinks are allowing weblogs link posts with one another. Comments are "reader-contributed replies to a specific post within the blog" (Marlow, 2004, p.3). The Comment system is implemented as a chronologically ordered set of responses and is the key form of information exchange in the blogosphere (Drezner & Farrell, 2004; Mishne & Glance 2006). "Posting volume would be a key determinant of content value" claimed Lu & Hsiao (2007, p. 346).

At last are trackbacks and pingbacks. Trackback is a citation notification system (Brady, 2005). It enables bloggers to determine when other bloggers have written another entry of their own that references their original post. "If both weblogs are enabled with trackback functionality, a reference from a post on weblog A to another post on weblog B will update the post on B to contain a back-reference to the post on A" (Marlow, 2004). A pingback is an automated trackback. "Pingbacks support auto-discovery where the software automatically finds out the links in a post, and automatically tries to pingback those URLs, while trackbacks must be done manually by entering the trackback URL that the trackback should be sent to" (http://codex.wordpress.org/ Introduction_to_Blogging#Pingbacks).

There are millions of individual blogs, but within any community, only a few blogs attract a large readership (Wagner & Bolloju, 2005). "The vast majority of blogs are probably only read by family and friends, there are only a few elite blogs which are read by comparably large numbers" wrote Jackson (2006, p.295). Herring et al. (2004) also claimed that most discussions of the blogosphere focus on an elite minority of blogs. These blogs are the most known and regularly linked by others. According to Trammell & Keshelashvili (2005, p. 968) their authors manage to create a persona, making themselves a "celebrity" among the community of bloggers. These blogs are referred as "A-list". "A-list blogs—those that are most widely read, cited in the mass media, and receive the most inbound links from other blogs—are predominantly filter-type blogs, often with a political focus. The A-list appears at the core of most characterizations of the blogosphere" wrote Herring et al. (2005).

Many bloggers desire a wide readership and the most reliable way to gain traffic to their blog is through a link on another weblog (Blood, 2002). Drezner & Farrell (2004, p.7) mentioned "when one blog links to another, the readers of the former blog are more likely to read the latter after having clicked on a hyperlink than they would have been otherwise. If they like what they read, they may even become regular readers of the second blog". In that way "blogs with large numbers of incoming links offer both a means of filtering interesting blog posts from less interesting ones, and a focal point at which bloggers with interesting posts, and potential readers of these posts can coordinate" (Drezner & Farrell, 2004; p.13). Less prominent bloggers contact one of the large 'focal point' blogs, to publicize their post when they have an interesting piece of information or point of view that is relevant to a political controversy. On the one hand this leads the readers to 'focal point' blogs, as they know that they will find links to many interesting stories, and on the other hand bloggers to send posts to focal point blogs as they know that they are likely to find more readers. Based on this and the lognormal distribution of weblogs, in a given a political issue, the media only needs to look at the top blogs to obtain a "summary statistic" about the distribution of opinions (Drezner & Farrell, 2004). In this vein Adamic & Glance (2005, p.2) noted: "Because of bloggers' ability to identify and frame breaking news, many mainstream media sources keep a close eye on the best known political blogs". Park & Jankofski (2008) investigated hyper linking of citizen blogs in South Korean politics and stated that "If there is an increasing frequency of neighbour links directly flowing through the blog of a politician, it may indicate the politician's role as the online community leader as well as the information hub for the community" (p.64). Drezner & Farrell (2004) found out that the median blogger has almost no political influence as measured by traffic or hyperlinks and they highlighted "This is because the distribution of weblinks and traffic is heavily skewed, with a few bloggers commanding most of the attention. This distribution parallels the one observed for political websites in general. Because of this distribution, a few "elite" blogs can operate as both an information aggregator and as a "summary statistic" for the blogoshpere". The same was claimed by Jackson (2006, p.296) who wrote "elite bloggers can act as a "focal point" encouraging influential visitors to congregate around them. To influence the news, political and policy agenda, political actors need to attract an "A" list audience to their weblog".

SOCIAL NETWORKING ANALYSIS

Social network analysis (SNA) is "a discipline of social science that seeks to explain social phenomena through a structural interpretation of human interaction both as a theory and a methodology" (Marlow, 2004 p.2). The goal of SNA is to identify "who the key actors are and what positions and actions are likely to take" (Krackhardt, 1996, p.161). According to Balancieri et al. (2007) SNA is rooted in the concepts of nodes and connections. Nodes, are the social actors and can be persons, groups, organizations, nations, communities, offices, blogs and so on and "connections" refer to channels of communication, (Balancieri et al., 2007; Martino & Spoto, 2006). Blogs facilitate members' social interactions and provide conversation (Nardi et al., 2004; Herring et al., 2005) and in this vein, SNA can be used as a research vehicle to investigate the structural patterns of blogging communities.

A Social Network can be represented in three ways: the first one is by giving a simple list of all the elements taken from the set of social actors, and the list of the pairs of elements that are linked by a social relationship of some kind. The second has a form of matrix. If two social actors I and J have a relation then 1 is placed at the cell (i,j), otherwise 0 is placed in this cell. Finally, a description of a Social Network may have a form of a graph where social actors can be represented by nodes, and the connections with each other can be represented by edges between these nodes. If the graph is directed, each interaction describes a one-way association between two social actors. In this case, the in-degree of a node is the number of incoming links and the out-degree is the number of outgoing links. Graph Theory's approach is crucial as it denotes the structural properties of a network and provides a tool to quantify and measure some properties of the network (Marlow, 2004; Martino & Spoto, 2006).

"Blogs are a form of social hypertext, functioning as a one-to-one mapping between a network of web pages and a network of people, which can be represented as a social network and from which communities can emerge" wrote Chin & Chignell (2007). Blog communities emerge from interlinking between them (Efimova & Hendrick, 2005; Efimova et al., 2005). SNA exploits interlinking between blogs and examines the roles and behaviour of nodes on other nodes in the network, and on the whole network (Chin & Chignell, 2007). Park & Jankowski (2008, p. 60) claimed "The configuration of link networks themselves can be a source conveying useful overall information about the (hidden) online relationship of communication networks in interpersonal, inter-organisational, and international settings"

Methodology

This research serves two purposes. On the one hand, it applies a hyperlink analysis of the incoming links of the recorded blogs using Social Networking Analysis (SNA), and on the other, it studies the communication patterns among blogs. It attempts to portrait the communication patterns generated among Greek political blogs through the linkages from blogrolls. The chapter records and analyzes blogroll hyperlinks interconnections. It uses

Technorati.com to track Greek political blogs and records blogs with tags to the five Greek parliamentary parties during May 2009. Technorati.com is considered a reliable and popular blog search engine (Chin & Chignell, 2007; Kolari et al., 2006; Zafiropoulos & Vrana, 2008).The five parliamentary Greek parties are: New Democracy (ND) - the Christian Democratic party which was in government at the time when this manuscript was prepared, Pan-Hellenic Socialist Movement (PASOK), which is the main opposition party, the Communist Party of Greece (KKE), Coalition of the Left and Progress (SYRIZA) and People's and the Orthodox's Rally (LAOS) a right wing party that has newly entered the parliament and is mainly characterized by nationalist and populist practices and rhetoric.

Through the search, 101 blogs with «some authority» were found. According to Technorati, authority is the number of blogs linking to a website in the last six months. The higher the number, the more Technorati Authority the blog has. In the authors view, this consideration grants greater validity concerning blog selection because the analysis takes into consideration only the blogs which are present and active for a while and probably are considered reliable.

Next, the chapter studies incoming links between blogs through their blogrolls. Originating from the Social Networking Theory, this study presents the blog connectivity patterns using directed graphs, where blogs are presented as nodes and incoming links between blogs are presented as arrows. The chapter presents hyperlinking patterns of the blogs using Social Networking Theory and graph theory. UCINET 6.0 for Windows is used to construct the network graph and to calculate graph theoretic indexes.

The directed graph, which presents the social network of blogs, is associated with its adjacency matrix. An adjacency matrix is a square non-symmetric binary data matrix where unity is placed in cell ij if blog i links to blog j through the blogroll, else zero is placed in the cell. Besides using Social Networking Theory and graph theoretic properties, the study goes a step further by applying multivariate statistical analysis to the data of the adjacency matrix. This research adopts a method introduced by Zafiropoulos & Vrana (2008) for locating central blog groups in political blogging. The original idea is that political blogs are organized around central focal point blogs, where most of the informative conversation is taking place (Drezner & Farrell, 2004). Zafiropoulos & Vrana (2008) introduced a combination of social networking theory, Multidimensional Scaling and Hierarchical Cluster Analysis to locate such groups by studying incoming links through blogrolls. By finding such groups, one can explore how bloggers are organized.

The multivariate statistical analysis consists of Multidimensional Scaling (MDS), which is used in the analysis as a data reduction technique on the one hand and to quantify the original binary data on the other. The method reproduces the original data and maps them on a fewer dimensions space (namely two in this analysis) while the effort is to keep intact the distances among the original data on the new reproduced data. "Stress" is a measure of goodness of fit between distances of original data and distances of the reproduced data. Better fit is assumed when stress is close to zero.

Next, Hierarchical Cluster Analysis (HCA) uses the quantified data from MDS to produce clusters of blogs, which have similar properties. Blogs in the same cluster are linked by nearly the same set of blogs. So in this way the blogs in a formed cluster are regarded to have common characteristics or be of the same family by blogs who link them. Some of the clusters that are produced by HCA gather the largest number of incoming links. If this happens then they may serve as conversational focal points. This property might be associated with the skewed distribution of links, mentioned by Drezner & Farrell (2004) for political blogs: only a few blogs have a very big number of incoming links, while the rest, the majority of blogs, have only a small number of incoming links. This formation of focal point blogs is not necessarily a direct consequence of the skewness of incoming links distribution since clusters can be located in any circumstances and scaling of the data is not a prerequisite. The chapter also presents the distribution of incoming blog links.

The study of the networks reveals how blogs are self-organized in groups around focal conversational points. Findings reveal whether the original hypothesis of Drezner & Farrell (2004) holds for the Greek case. Further, by exploring inter-cluster linkages, the analysis explores whether Greek political blogs are characterized by a certain degree of polarization. The study of polarization may interpret whether blog clusters political affiliation is associated to blogs interconnections.

FINDINGS

Part 1: A Social Networking Theory Approach

Links distribution

In this section, the chapter follows a Social Networking Theory (SNT) approach. Incoming and outgoing links distributions and statistics give a sense of the size of blog interconnectivity (Table 1). The number of outgoing links ranges from zero to 27 with a mean value 4.87 and a standard deviation which is greater than the mean (5.56). Dispersion of the number of incoming links is a little bit smaller since maximum value is 16 and standard deviation is 3.46 while mean value is 4.87 as it is for incoming links.

Figure 1 shows the graphic representation of the links distributions. Distributions appear skewed. Skewness is high for both distributions, however skewness of outgoing links is much higher that that of the incoming links. Skewness denotes the situation where only a few blogs have the biggest number of links, while the majority has only a few links. Therefore, from the analysis it is obvious that this property holds for both incoming and outgoing links, but it is more intense for outgoing links.

Centrality

The centrality of a node in a network is a measure of the structural importance of the node. A person's centrality in a social network affects the opportunities and constraints that they face. In this chapter, we consider two forms of centrality: density, and betweenness.

Table 1. Descriptive Statistics of the links

	N	Minimum	Maximum	Mean	Std. Deviation	Skewness
Outgoing links	101	.00	27.00	4.8713	5.56896	1.806
Incoming links	101	.00	16.00	4.8713	3.46024	1.008

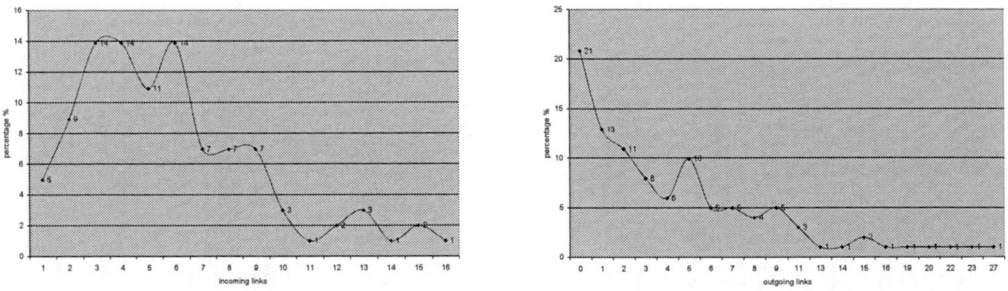

Figure 1. A detailed presentation of the distribution of blogs incoming and outgoing links

Density: Degree is simply the number of nodes that a given node is connected to. In general, the greater a person's degree, the more potential influence they have on the network, and vice-versa. For example, in a community network, a person who has more connections can spread information more quickly, and will also be more likely to hear more stuff. The greater a person's degree, the greater the chance that they will catch whatever is flowing through the network. The "density" of a binary network is the total number of ties divided by the total number of possible ties. In the case of Greek blogs, density (matrix average) equals 0.0485 or 4.85% with standard deviation 0.2149. It becomes obvious that density is extremely low. The network as a whole seems lacking interconnections or there exist just few sparse links between some blogs.

Betweenness: Loosely speaking, betweenness centrality is defined as the number of geodesic paths that pass through a node. It is the number of "times" that any node needs go through a given node to reach any other node by the shortest path. The node with high betweenness can serve as a liaison between disparate regions of the network. Betweenness is therefore a measure of the number of times a vertex occurs on a geodesic. The normalized betweenness centrality is the betweenness divided by the maximum possible betweenness expressed as a percentage. In this specific blogs' network mean betweenness equals 1.555 and its standard deviation equals 2.164. There is a lot of variation of normalized betweenness since standard deviation is much higher than the mean. This means that there exist some blogs that have the property of betweenness, while on the other hand there are blogs that do not have this property. Further analysis of the distribution of normalized betweennes for the 101 blogs, reveals that 74% of them have the property of beweenness, while the rest 26% of the blogs do not have this property.

Network Centralization Index can be regarded as a measure for the network to have central points or areas with greater number of paths than usual. Network Centralization Index equals 9.14%. Despite big variance of normalized betweenness, this index is low. In conclusion, the specific network has only few central blogs (concerning their links) and they do not differ significantly from the rest regarding the property of centrality and the property of linking to others.

Components and Cliques

Cliques and components are sets of blogs that are connected in certain ways. Finding cliques is an attempt to understand whether blogs interact as being linked friends in a social network.

Components: A connected component is a maximal subgraph in which all nodes are reachable from every other. In a directed graph, two vertices are in the same weak component if there is a semi-path connecting them. In this directed network, two components were found. Regarding the components sizes, one component includes most of the blogs: 99% in one component and 1% in the other. It can be calculated that fragmentation, that is the proportion of blogs that cannot reach each other, equals 2%. This number is considered very small, and it probably means that in general, one user can navigate through these blogs using links from blogrolls, and practically with some effort and time spent; he/she could visit nearly all the blogs. However, things look a little different when studying cliques.

Cliques: For the needs of the analysis, the original adjacency matrix is symmetrised. This means we consider that two blogs are connected if either one links to the other. A clique is a maximally complete subgraph. Maximal means that it is the largest possible subgraph: you could not find another node anywhere in the graph such that it could be added to the subgraph and all the nodes in the subgraph would still be connected. A complete graph is a simple graph in which every pair of distinct nodes-vertices is connected by an edge.

The number of cliques and the number of blogs that form cliques could be regarded as a measure of the volume of interconnectivity of the network. It shows how blogs are connected to form small companies of friends. 88% of the blogs form 193 cliques: 7 cliques (3.6%) with 5 blogs, 42 (21.8%) with 4 blogs, and 144 (21.8%) cliques with 3 blogs.

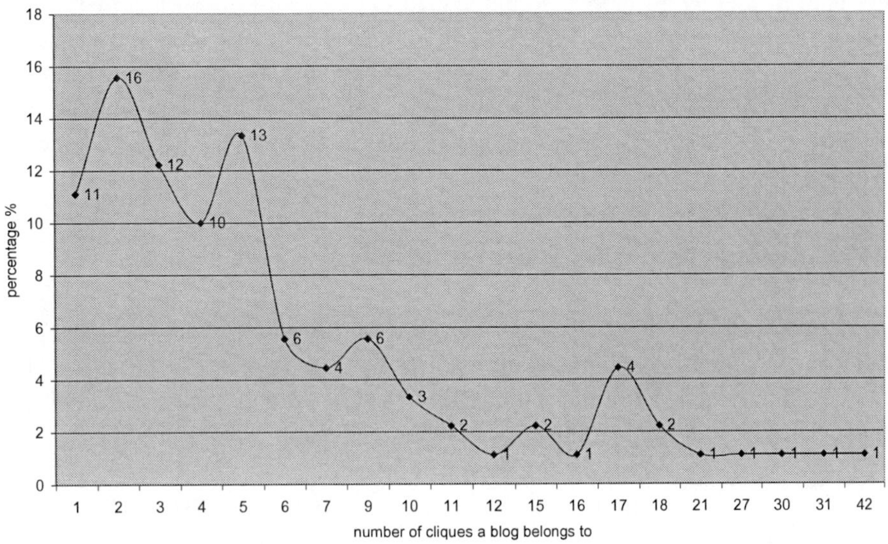

Figure 2. Distribution of blogs according to incorporation to cliques

The line graph in Figure 2 presents how many blogs take part in several cliques at the same time. This is a measure of popularity and a measure of interconnection at the same time. Over 60% of the blogs in cliques are member of one to fine cliques.

Concluding, we can say that group formation is high for Greek blogs. Links are few, however there is almost no isolated blogs, while groupings of blogs are common.

Co-Citation

This section studies the way that blogs are co-cited by other blogs, that is, how they are linked simultaneously by more than one blog. Co-citations are used in the sense that they may be indexes of popularity of the blogs. Alternatively, they may be indexes of describing to what degree do blogs recognize central – core groups of other blogs and in this way, they point to them. This is an important concept since it deals with the property of the networks self-realization, that is, the way that blogs have a view of what happens in the network, who links to whom, who are the most interesting blogs, etc. Let us revisit the example from the study of political blogging: within polarized political systems blogs are forming clusters around central blogs, which are considered reliable or having the same affiliation. Users of the Internet who wish to be informed quickly locate the focal points of discussion and for economy of navigation, they read only the posts on these blogs. Bloggers also locate focal point blogs and place their posts along with a link to their blog. They expect thus that the readers of focal point blogs will also visit their blogs (Vrana & Zafiropoulos 2009, Zafiropoulos & Vrana 2008).

Co-citations are calculated by multiplying the transposed adjacency matrix by the original adjacency matrix. The product is a symmetric binary matrix. Its elements in cell ij is the number of blogs that link both blogs i and j. Therefore, for each blog a series of 101 numbers is produced: it presents the numbers of co-citations with any other blog. Obviously, it is very hard to present all these findings. For economy, we use just a part of this data set. For each blog, the maximum co-citation with any other blog is chosen and is presented in Figure 3. For example, 5% of the blogs are not co-cited with any other blog, while 9% of the blogs are co-cited along with one other blog. More than 50% of the blogs are co-cited by 2, 3, 4 or 5 blogs through blogrolls. There is quite enough co-citation within political blogs.

Part 2: Exploring Communication Patterns

Drezner & Farrell (2004, p.13) mention “Blogs with large numbers of incoming links offer both a means of filtering interesting blog posts from less interesting ones, and a focal point at which bloggers with interesting posts and potential readers of these posts can coordinate. When less prominent bloggers have an interesting piece of information or point of view that is relevant to a political controversy, they will usually post this on their own blogs. However, they will also often have an incentive to contact one of the large ‘focal point’ blogs, to publicize their post. The latter may post on the issue with a hyperlink back to the original blog, if the story or point of view is interesting enough, so that the originator of the piece of information receives more readers. In this manner, bloggers with fewer links function as “fire alarms" for focal point blogs, providing new information and links. Also they mention that: "We note that this implies that even while focal point blogs play a crucial mediating role, smaller blogs may sometimes have very substantial political impact by bringing information to the attention of focal blogs" (Drezner & Farrell, 2004; p.13). This chapter argues that "focal point" blogs are recognized as authority blogs by the blogging community and they may serve as the blog cores where the interesting and informational discussion is taking place. This property can be used to limit the analysis only to these blogs, excluding in this way other blogs which might be considered to have limited interconnections with other blogs or they are isolated. This property might be a consequence of the skewed distribution of links, also mentioned by Drezner & Farrell (2004).

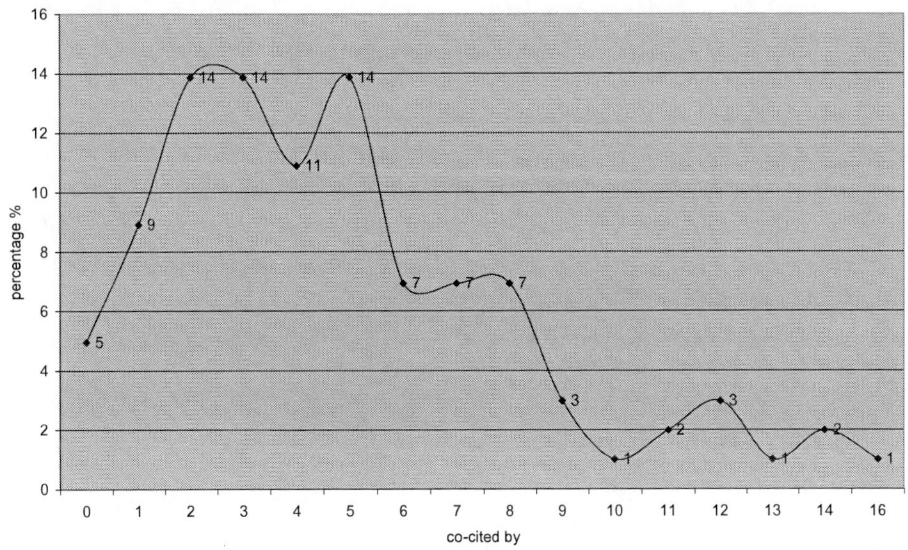

Figure 3. Distribution of blogs proportions according to co-citations

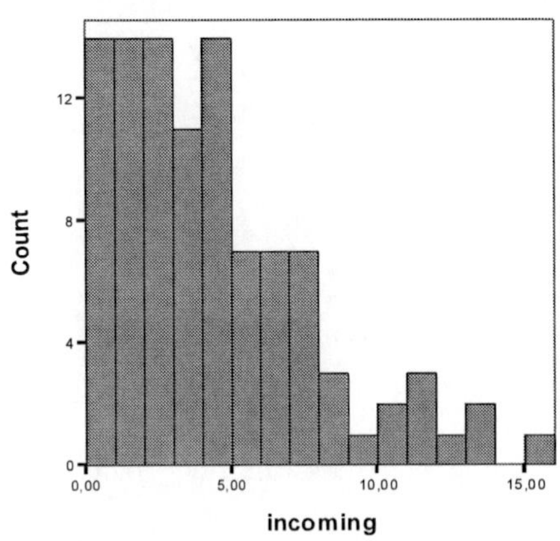

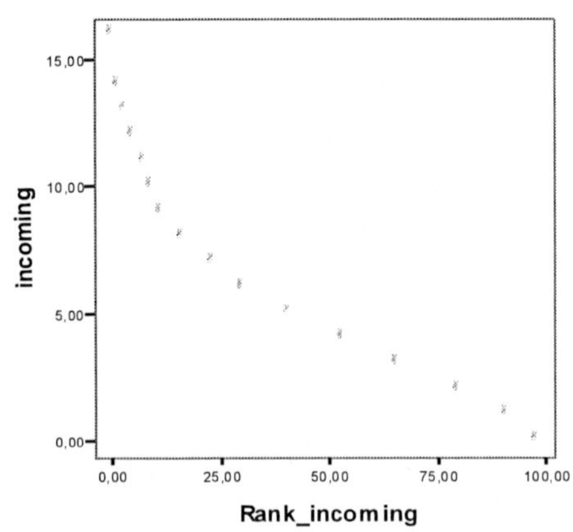

Skewness=1.008

Figure 4. Histogram of blogs incoming links (left) and scaterplot of blogs ranks according to incoming links vs actual number of incoming links-indegrees(right)

To test whether this hypothesis holds for political blogging in Greece, this study examines the distribution of incoming links to the 101 blogs of the study. Figure 4 presents blog ranks according to incoming links (1 being the 1st blog in order, that is the blog linked the most) versus the percentage of incoming links. From these figures, it is obvious that there exists a skewed distribution, which implies that only a few blogs gather the highest number of incoming links. This finding provides evidence that Drezner & Farrell's (2004) argument about the skewedness of incoming links distribution holds true. Similar findings are produced when outgoing links are studied. Outgoing links are links from blog blogrolls to other blogs. They present the way that bloggers connect to other bloggers and they may reveal connectivity patterns between "neighbouring" bloggers. Skewness of outgoing links distribution is even greater than that of incoming links, and this finding shows that only few bloggers are really engaged in connecting and pointing to others while the rest of (the majority) are lining to only a few or even no other blogs (Figure 5).

To analyze this situation further and to locate core blog groups, statistical analysis using Multidimensional Scaling (MS presents a very good fit with Stress=0.07), followed by Hierarchical Cluster Analysis (HCA) is performed. Using a scree-plot of a number of clusters versus Wilk's Lambdas, it seems that the optimal number of blogs is five. Five clusters of blogs are formed regarding the incoming links. These clusters of blogs are described in Figure 6.

Formation of groups is a finding which is also compatible with those founded by Drezner & Farrell (2004). When these clusters are studied in relation to their size and number of interconnections, then some interesting results may arise.

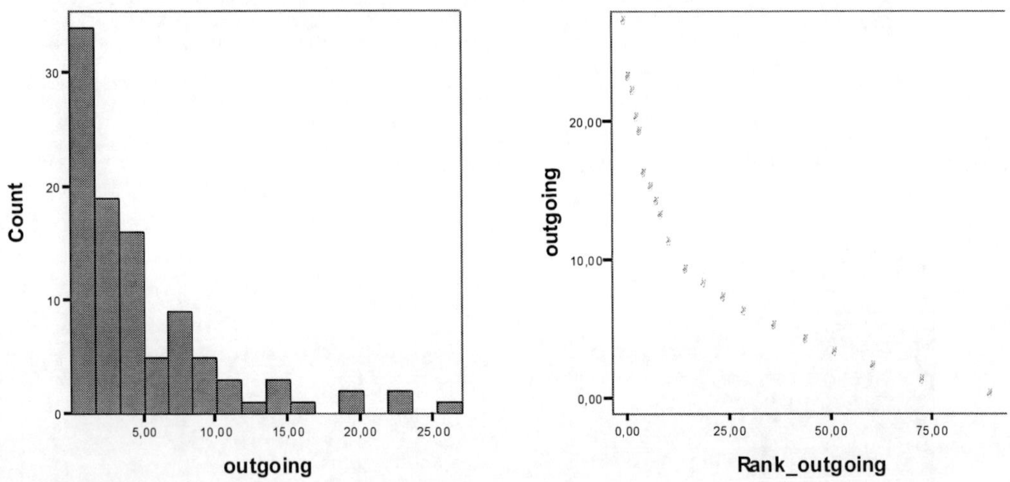

Skewness=1.806

Figure 5. Histogram of blogs outgoing links (left) and scatterplot of blogs ranks according to outgoing links vs actual number of outgoing links-outdegrees (right)

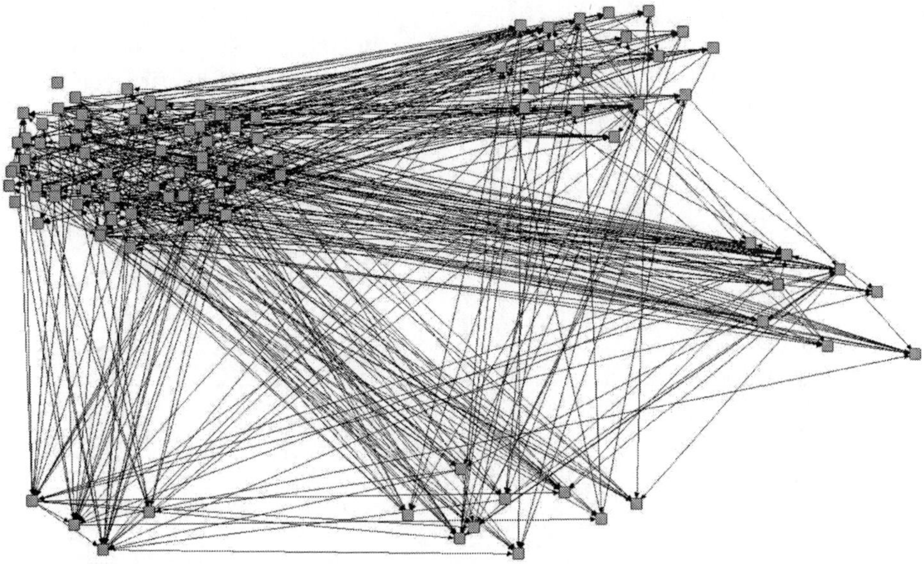

Figure 6. Blogs Social Network according to incoming links through blogrolls. Cluster 1 is placed at the upper left side of the graph, while clusters 2 to 5 are placed clockwise.

DESCRIPTION OF THE CLUSTERS

Statistics and Political - Ideological Affiliation

Table 2 presents the five clusters in descending order of size from the left to the right. Cluster 1 for example contains more than 60% of the blogs. Although it is the most populous cluster, it presents the lowest rate of both incoming and outgoing links. Thus, the cluster contains the least active blogs regarding networking. Regarding the other clusters, there is strong negative correlation between cluster size and percentage of incoming links. The more we move towards clusters 5, the more the average incoming link rate is and the lower the size of the cluster becomes. This finding is summarized in Figure 7 where the negative correlation of the percentage of blogs in every cluster and the percentage of incoming links is obvious. Leaving out cluster 1, we could limit our analysis to and comment on the remaining four clusters, those that have many incoming links and small size at the same time. In particular clusters 3, 4, and 5 on average receive more than 9 incoming links. Because the recorded blogs are 101, this number can be directly interpreted as being a percentage of nearly 9%. Further, the most linked cluster is consists of just four blogs and has an average of 10.7 of incoming links.

Regarding the tags of the blogs within each cluster, cluster 2 presents a high rate of tagging to the Communist Party of Greece (KKE) and SYRIZA, since 72.2% of the blogs within the clusters have tags to KKE while 50% of the blogs tag to SYRIZA. It is a group of blogs maintained mainly by leftists with references to KKE and SYRIZA. Cluster 3 tags mainly PASOK while it keeps tagging to the other parties very low. Blogs in this cluster are maintained by members and friends of PASOK. Clusters 4 and 5 have high rates of tags to

PASOK, ND and SYRIZA, while 50% of blogs in cluster 5 tag to KKE. These clusters are formed of blogs, which provide information argumentation, and speculation about society, and digital liberties (Table 3). It is very interesting to note that that the frequency of blogs with a Right wing affiliation is almost null (Table 3).

Table 2. Clusters properties

Clusters	1	2	3	4	5
Frequency	62	18	8	9	4
Percentage	61.4%	17.8%	7.9%	8.9%	4.0%
Outgoing links (average)	3.0	6.6	9.8	9.7	4.7
Incoming links (average)	2.8	6.2	9.3	9.5	10.7
Tags to PASOK	56.5%	44.4%	75.0%	66.7%	75.0%
Tags to ND	58.1%	44.4%	37.5%	77.8%	75.0%
Tags to SYRIZA	35.5%	50.0%	12.5%	77.8%	75.0%
Tags to KKE	32.3%	72.2%	25.0%	33.3%	50.0%
Tags to LAOS	24.2%	22.2%	12.5%	33.3%	25.0%

Table 3. Clusters' profiles and affiliation. Clusters sorted in ascending order of average number of incoming links

Cluster 2	Mainly Left, KKE and SYRIZA
Cluster 3	PASOK
Cluster 4	Left with broader speculation about society and democracy
Cluster 5	Left, digital liberties, information provision, discussion and argumentation

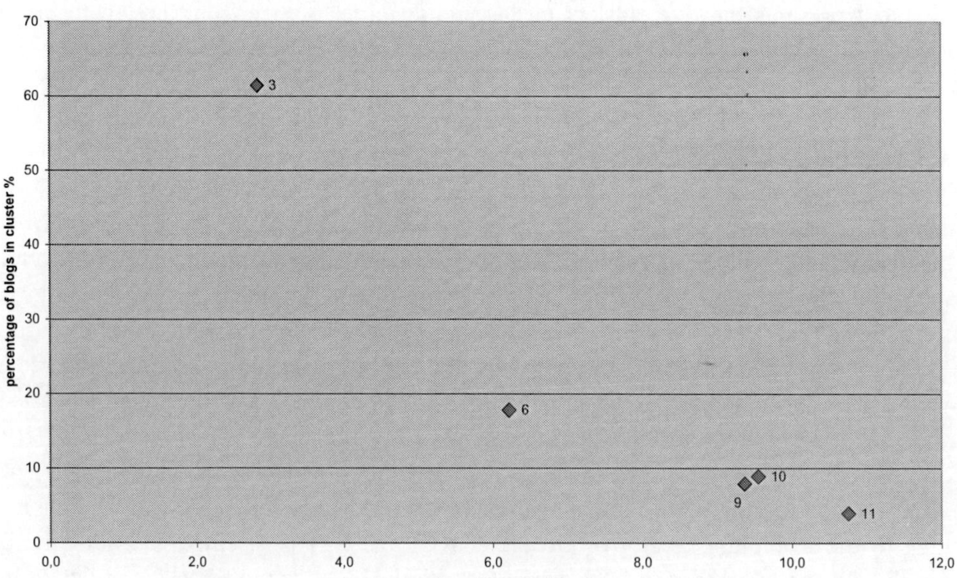

Figure 7. Scatterplot of clusters according to their characteristics: incoming links-number of blogs within clusters

Measuring Partisanship and Polarization Using the Clusters Internal Coherence and External Divergence

Since clusters of blogs are to a certain degree affiliated to a political party or to some political orientation, it is important to explore 1) how strong the internal communication between blogs of the same cluster is. To answer this question we measure the number of incoming links among blogs within each cluster. Next, we explore 2) how strong the external divergence between clusters is. To answer this question we proceed by counting the incoming links among blogs of different clusters. In a polarized system, we expect intense communication among blogs of the same clusters or clusters of familiar affiliation on the one hand and weak communication among blogs of different clusters. Throughout this chapter, the number of links among blogs were used to meansure the strength of communication. In this particular case, polarization should be noticed when links within clusters are many while links among between clusters are few.

A cluster is considered partisan if all its blog members are of the same affiliation and they are interconnected to a high degree within the cluster. The particular political blogosphere is considered polarized if clusters are hardly interconnected while there is high enough interconnections within the clusters. In a polarized system there are only few incomings links from blogs of a cluster to the blogs of any other clusters, while there are a lot of links within each cluster.

Internal coherence of a cluster of blogs refers to the degree that the blogs within the cluster link to each other. In other words, when many blogs within the cluster connect to each other, internal coherence is high. Therefore, internal coherence maximum value equals the maximum number of incoming links within the cluster, and this equals $n \cdot (n-1)$ where n stands for the number of blogs within the cluster.

External divergence emerges when there are only a few links between two clusters of blogs. If this happens then there are only sparse links from blogs of one cluster to blogs of the other. The maximum number of links between two clusters equals $n \cdot m$ where n stands for the number of clusters for the one cluster and m the number of the clusters of the other.

To calculate indexes to describe coherence and divergence we work as follows:

At first, a 5 by 5 square matrix, matrix A, is formed that contains the maximum number of links between blogs, as described above. Thus in cell i,j of this matrix, the maximum number of links that could appear joining cluster i with cluster j is calculated.

Next, the actual number of links between cluster i and cluster j are calculated and are recorded in a matrix B. Cell i,j of this matrix contains the number of incoming links from cluster i to cluster j as they are recorded in the present research.

At the third step, cell i,j of matrix B is divided by cell i,j of matrix A and the outcome is presented as a percentage. This is an index of interconnection degree between clusters of blogs. The diagonal elements of this final product matrix should be high in order that the clusters to have high internal coherence, while the rest of the cells should have small values in order that the clusters have a low degree of connectivity i.e. a high degree of divergence.

Table 4 contains the outcomes of these calculations. The number in any cell i,j denotes the percentage of incoming links to cluster j from cluster i. Cluster 1 appears to be the least active regarding connections. Blogs in cluster 1 neither receive nor do they send any links to blogs of the other clusters. Furthermore, internal coherence for this cluster is very low, since only 2% of all the available links are really used. For the rest of the clusters internal coherence is quite good. It nearly equals 18% except for cluster 3 (affiliated to PASOK) where internal coherence rises to 34% being double that of the internal coherence of each one of the clusters 2, 4, and 5. Having a very clear political affiliation to Pan-Hellenic Socialist Movement and a high degree of internal coherence, cluster 3 seems to be the most partisan cluster. It consists of very active members who communicate and connect to each other. In addition, cluster 3 has high links percentages to and from clusters 4 and 5.

Cluster 2 has a good degree of internal coherence (18%) and very low degrees of external connectivity to and from the other clusters. However, links percentages to clusters 4 and 5 both equal 11% and incoming links percentage from cluster 4 to cluster 2 equals 12%. It is a more isolated cluster of blogs with a fair internal coherence. Taking into consideration that this cluster contains blogs affiliated mostly to KKE and SYRIZA, we could define it as partisan with few ties to other clusters.

Cluster 4 presents a good degree of internal coherence at 18% and good ties (links percentages) to clusters 2, 3, and 5. Its links percentages to and from cluster 5 are very high, 25% and 22% respectively. Clusters 4 and 5 are really well connected. This property might be associated with the fact that they share common believes and ideas.

Table 4. Percentages of incoming links with respect to the maximum number of incoming links between clusters

	Cluster 1	Cluster 2	Cluster 3	Cluster 4	Cluster 5
Cluster 1	2%	3%	7%	6%	8%
Cluster 2	3%	18%	4%	11%	11%
Cluster 3	8%	2%	34%	18%	16%
Cluster 4	7%	12%	13%	18%	25%
Cluster 5	2%	1%	13%	22%	17%

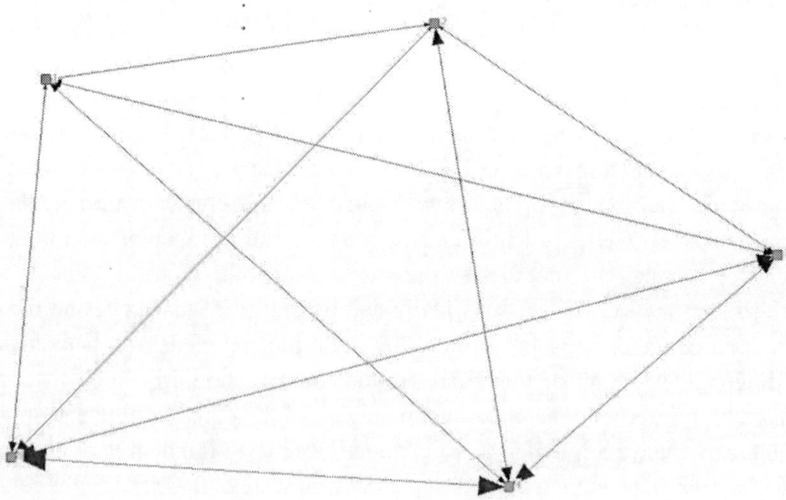

Figure 8. Strength of interconnections between clusters of blogs

Cluster 5 looks very similar to cluster 4, regarding percentages of incoming links. However, it differs from cluster 4 in the sense that while cluster 2 links fairly well to it, cluster 5 does not link to cluster 2 (1%). Blogs in cluster 2 might regard blogs of cluster 5 as interesting or being politically close to them, but the opposite does not seem to hold.

In conclusion, clusters 4 and 5 present the highest degrees of interconnections or ties. They both are fairly well connected to cluster 3, but compared to it they have half of its internal coherence. Cluster 3 is the cluster with the most active and interconnected members. Clusters 3, 4 and 5 form a larger extended cluster of smaller connected clusters. Cluster 2 has more sparse connections to clusters 3, 4, and 5 and seems to be isolated from the other clusters, but contains active members. Figure 8 presents the strength of interconnections between different clusters of blogs. Larger arrowheads are used to present higher percentages of incoming links. The block of clusters 3, 4, and 5 is presented at the lower part of the graph. The connections between these clusters are denoted with larger arrowheads.

CONCLUSIONS

The chapter made a broad hyperlink analysis of Greek political blogs. Hyperlinks are important to bloggers not only to maintain some communication with each other, but to determine their digital territory as well. Regarding the social networking analysis of the blogs, hyperlinks are few, yet connectivity is high. This means that in general, there is always a path (through blog hyperlinks) that can lead a user from a blog to another blog, simply by following consecutive links. In this sense, isolated blogs hardly exist. In addition, co-citation is high. Blogs are linked by groups of other blogs. This supports the hypothesis that bloggers manage to locate central blogs groups and they gather around them linking to them. They form, in this way, groups of blogs around central and significant clusters of blogs, because they find that these central clusters of blogs are of the same affiliation or they are reliable or informative.

A portion of bloggers are self-organized in central clusters of blogs having the same political or ideological affiliation, and to a certain degree, they develop interconnections within each cluster. The degree of interconnections varies across clusters. Some clusters present high degrees of internal coherence, but in any case, internal coherence for every such cluster is big. In addition, connection between clusters exists. In this way, this system of clusters of blogs reveals that blogs, most of them having a left-wing orientation, are organized in a system, which is partisan but not much polarized. Further, when looking at the big picture, it becomes clear that the basic hypothesis of focal conversational points creation supported in the literature is valid to a certain degree for the Greek case too. Blogs are organized in groups around clusters of central blogs which share a common affiliation. These core groups can be located using multivariate statistical techniques presented in this chapter. Having these properties, the Greek political blogosphere, although limited, seems to have grown from the early stage phase to the phase of self-organization and self-realization.

References

Adamic, L. & Glance, N. (2005). The Political Blogosphere and the 2004 U.S. Election: Divided They Blog. In proceedings of the WWW-2005, workshop on the Weblogging Ecosystem. http://www.blogpulse.com/papers/ 2005/ AdamicGlanceBlogWWW.pdf.

Albrecht, S., Lübcke, M. & Hartig-Perschke, R. (2007). Weblog Campaigning in the German Bundestag Election 2005. *Social Science Computer Review* , *25(4)*, 504-520. Allan, S. (2002). Reweaving the Internet: Online news of September 11. In: B. Zelizer, & S. Allan, (Eds.), *Journalism after September 11* (119-140). New York: Routledge.

Andrews P. (2003). Is blogging journalism? *Nieman Reports*, *57(3)*, 63-64. http://download.101com.com/syllabus/conf/summer2004/PDFs/w01.pdf.

Balancieri, R., Cuel, R. & dos Santos Pacheco, R. C. (2007). Social Network Analysis for Innovation and Coordination. Proceedings of I-KNOW '07 Graz, Austria, September 5-7, 2007; http://triple-i.tugraz.at/blog/wp-content/uploads /2008/ 11/9_social-network-analysis-for-innovation-and-coordiniation.pdf.

Baker, S. & Green, H. (2005). Blogs will change your business. *Business Week*, 5 February, 3931; 56-65.

Banning, S. & Trammell, K. D. (2006). Revisiting the issue of blog credibility: A national survey. Paper presented at the annual conference at the Association for Education in Journalism & Mass Communication, August, San Francisco, CA.

Barger, J. (1997). *FAQ: Weblog Resources*. http://www.robotwisdom.com/weblogs/

Bar-Ilan, J. (2007). The use of Weblogs (blogs) by librarians and libraries to disseminate information. *Information Research*, *12(4)*, paper 323. http://informationr.net/ir/12-4/paper323.html (6/4/2007).

Betts, D. J. & Glogoff, S. J. (2004). Instructional models for using weblogs in elearning: A case study from a virtual and hybrid course. *Syllabus 2004*. http://chnm.gmu.edu/ digitalhistory/links/pdf/chapter1/1.41.pdf.

Bloom, J. (2003). The blogosphere: how a once-humble medium came to drive elite media discourse and influence public policy and elections. Paper presented to the American Political Science Association, Philadelphia, PA, August.

Bloom, J. & Kerbel, M. (2005). *The blogosphere and the 2004 elections*. Paper presented at the American Political Science Association Annual Conference, Washington, DC, September.

Blood, R. (2002). *The Weblog Handbook: Practical Advice on Creating and Maintaining Your Blog*. Cambridge MA: Perseus Publishing.

Brady, M. (2005). Blogging: personal participation in public knowledge building on the web Chimera Working Paper Number: 2005, 02. www.essex.ac.uk /chimera/content/ pubs/ wps/cwp-2005-02-blogging-in-the-knowledge- society-mb.pdf.

Chadwick, A.(2008). Web 2.0: New Challenges for the Study of E-Democracy in an Era of Informational Exuberance. *I/S A Journal of Law and Policy for the Information Society, 4 (3)*,10-42.

Chin, A. & Chignell, M. (2007). Identifying communities in blogs: roles for social network analysis and survey instruments. *Int. J. Web Based Communities*, *3(3)*, 345-363.

Cornfield, M., Carson, J., Kalis, A. & Simon, E. (2005). *Buzz, blogs and beyond.* Pew Internet & American Life Project. www.pewinternet.org/ppt/buzz_blogs_beyond_final05-16-05.pdf.

Dwivedi, M., Shibu. T. P. & Venkatesh, U. (2007). Social software practices on the internet. Implications for the hotel industry. *International Journal of Contemporary Hospitality Management, 19(5)*, 415-426.

Drezner, D. & Farrell, H. (2004). The power and politics of blogs, paper presented at the Annual Meeting of the American Political Science Association, Washington, DC, August.http://www.utsc.utoronto.ca/~farrell/ blogpaperfinal.pdf.

Du, H. & Wagner, C. (2006). Weblog success: Exploring the role of technology. International. *Journal of Human-Computer Studies*, *64*, 789-798.

Efimova, L. & Hendrick, S. (2005). In search for a virtual settlement: an exploration of Weblog community boundaries. https://doc.novay.nl/dsweb/Get/Document-46041.

Efimova, L., Hendrick, S. & Anjewierden, A. (2005). Finding "the life between buildings": an approach for defining a weblog community. AOIR Internet Research 6.0: Internet Generations, Chicago. https://doc.novay.nl/dsweb/Get/Document-55092/AOIR_blog_communities.pdf.

Festa, P. (2003). Blogging comes to Harvard. Newsmaker. http://news.com.com/2008-1082-985714.html?tag=fd_nc_1.

Fuchs, C. (2007). Towards a dynamic theory of virtual communities. *International Journal of Knowledge and Learning, 3(4/5)*, 372 - 403.

Garrett, J. (2004). User experience analysis: Presidential campaign sites. Report prepared for Adaptive Path.

Glover, D. (2006). The Rise of Blogs. National Journal, January 21, 2006, republished online at Daniel Glover's Beltway Blogroll blog http://beltwayblogroll.nationaljournal.com/archives/2006/01/the_rise_of_blo.php.

Graf, J. (2006). The Audience for Political Blogs. New research on Blog Readership.GW's Institute for Politics, Democracy & the Internet. George Washington University, Washington, DC http://www.ipdi.org/uploadedfiles/audience %20for%20political%20blogs.pdf.

Graf , J. & Darr, C. (2004). Political Influentials Online in the 2004 Presidential Campaign, Institute for Politics, Democracy and the Internet, George Washington University, Washington, DC, February, http://www.ipdi.org/UploadedFiles/ political%20influentials . pdf.

Gumbrecht, M. (2004). Blogs as "protected space" WWW 2004. Workshop on the Weblogging Ecosystem: Aggregation, Analysis and Dynamics.

Guo, X., Vogel, D., Zhou, Z., Zhang, X. & Chen, H. (2008). Chaos Theory as a Model for Interpreting Weblog Traffic. Proceedings of the 41st Hawaii International Conference on System Sciences – 2008.

Hepburn, C. (2007). web 2.0 for the tourism and travel industry. http://www. ebusinessforum.gr/engine/index.php?op=modload&modname=Downloads&action=downloadsviewfile&ctn...el.

Herring, S. C., Scheidt, L. A., Bonus, S. & Wright, E. (2004). *Bridging the gap: A genre analysis of weblogs.* Proceedings 37th Annual HICSS Conference 2004. Big Island, Hawaii.

Herring, C., Kouper, I.Paolillo, J., Scheidt, L. A.,Tyworth, M., Welsch, P., Wright, E. & Yu, N. (2005). *Conversations in the Blogosphere: An Analysis "From the Bottom Up"* Proceedings of the Thirty-Eighth Hawai'i International Conference on System Sciences (HICSS-38). Los Alamitos.

Hiler, J. (2002). *Blogs as disruptive tech: How weblogs are flying under the radar of the content management giants*. http://www.webcrimson.com/ourstories/ blogsdisruptivetech. htm.

Hoogenboom, T., Kloos, M., Bouman, W. & Jansen, R. (2007). Sociality and learning in social software. *International Journal of Knowledge and Learning*, *3(4/5)*, 501-514.

Hsu, C. L. & Lin, C. C. (2008). Acceptance of blog usage: The roles of technology acceptance, social influence and knowledge sharing motivation. *Information and Management, 45*, 65-74.

Jackson, N. (2006). Dipping their big toe into the blogoshpere. The use of weblogs by the political parties in the 2005 general election. *Aslib Proceedings: New Information Perspectives*, *58(4)*, 292-303.

Johnson, T. & Kaye, B. (2004). Wag the blog: how reliance on traditional media and the internet influence credibility perceptions of weblogs among web users. *J&MC Quarterly*, *81(3)*, 622-642.

Johnson, T. J. & Kaye, B. K. (2007). Choosing is believing? How Web gratifications and reliance affect Internet credibility among politically interested users. Paper presented to the Association for Education in Journalism and Mass Communication, August, Washington, D.C.

Johnson, T. J., Kaye, B. K., Bichard, S. L. & Wong, W. J. (2007). Every blog has its day: Politically-interested Internet users' perceptions of blog credibility. *Journal of Computer-Mediated Communication*, 13(1), paper 6. http://jcmc.indiana.edu/vol13/issue1/ johnson.html.

Jüch, C. & Stobbe, A. (2005). Blogs: The new magic formula for corporate communications? Economics. Digital economy and structural change. *Deutsche Bank Research*, *53*, 22 August.

Kahn, R. & Kellner., D.(2004). New media and Internet activism: From the "Battle of Seattle" to blogging. *New Media & Society*, *6(1)*, 87-95.

Kerbel, M. R. & Bloom, J. D. (2005). Blog for America and civic involvement. *The Harvard International Journal of Press/Politics*, *10*, 3-27.

Kim, H. N. (2008). The phenomenon of blogs and theoretical model of blog use in educational contexts. *Computers & Education*, *5*, 1342-1352.

Koh, A., Lim, A., Soon, Mg-Ee., Detenber, B. & Cenite, M. (2005). Ethics in Blogging. Singapore Internet Research Centre. http://www.ntu.edu.sg/sci/sirc/.

Kolari, P., Finin, T. & Joshi, A. (2006) SVMs for the Blogosphere: Blog Identification and Splog Detection. In: AAAI Spring Symposium on Computational Approaches to Analysing Weblogs. Maryland: University of Maryland.

Kolbitsch, J. & Mauer, H. (2007). The Growing Importance of e-Communities. *Web Lecture Notes in Business Information Processing, 1*, 19-37.

Kosonen, M., Henttonen, K. & Ellonen, H-K.(2007). Weblogs and internal communication in a corporate environment: a case from the ICT industry. *International Journal of Knowledge and Learning*, *3(4/5)*, 437-449.

Krackhardt, D. (1996). Social Networks and the Liability of Newness for Managers. *Trends in Organizational Behavior*. In Cooper, C.L. and Rousseau D. M. (eds.), (pp. 159-173). John Wiley and Sons, Ltd. New York, NY.

Krishnamurthy, S. (2002). The Multidimensionality of Blog Conversations: The Virtual Enactment of September 11. In Maastricht, The Netherlands: *Internet Research*, 3.0.

Lankshear, C. & Knobel M. (2003). Do-It-Yourself Broadcasting: Writing Weblogs in a Knowledge Society Paper presented to the American Education Research Association Annual Meeting. Chicago, April 21 http:// www.ballarat.edu.au.

Lasica, J. D. (2001). Blogging as a form of journalism. USC Annenberg. *Online Journalism Review*. http://www.ojr.org/ojr/workplace/1017958873.php.

Lasica, J. D. (2002). Weblogs: A new source of news. In R. Blood (Ed.), *We've Got Blog: How Weblogs are Changing Our Culture* (pp. 171-182). Cambridge, MA: Perseus Publishing.

Lu, H. P. & Hsiao, K. L. (2007). Understanding intention to continuously share information on weblogs. *Internet Research*, *17(4)*, 345-361.

Maag, M. (2005). The potential use of ‘‘Blogs” in nursing education. *CIN: Computers, Informatics, Nursing*, *23*, 16-24.

Marlow, C. (2004). Audience, structure and authority in the weblog community. The 54th Annual Conference of the International Communication Association, http://www. researchmethods.org/ICA2004.pdf.

Martino, F. & Spoto, A. (2006). Social Network Analysis: A brief theoretical review and further perspectives in the study of Information Technology. *PsychNology Journal*, *4(1)*, 53-86.

McKenna, L & Pole, A. (2003). Do Blogs Matter? Weblogs in American Politics American *Political Science Association, Annual Meeting Chicago*, IL, September 2004.

Mishne, G. & Glance, N. (2006). Leave a reply: An analysis of weblog comments. WWW May, 22-26, 2006, Edinburgh, UK. http://www.blogpulse.com/ www2006-workshop/papers/wwe2006-blogcomments.pdf.

Nardi, B. Schiano, D. & Gumbrecht, M. (2004). ‘Blogging as social activity, or, would you let 900 million people read your diary?’, Proceedings of the 2004 ACM Conference on Computer Supported Cooperative Work, New York, USA, 222-231.

Park, H. W. & Jankowski, N. (2008). A hyperlink network analysis of citizen blogs in South Korean Politics. Javn*ost-the Public*, *15(2)*, 57-74.

Pedley, P. (2005). International phenomenon? Amateur journalism? Legal minefield? Why information professionals cannot afford to ignore weblogs. *Business Information Review*, *22(2)*, 95-100.

Pohlig, C. (2003) How a Newspaper Becomes H.I.P. *Neiman Reports* (Winter): 24-6. http://www.nieman.harvard.edu/reportsitem.aspx?id=100906.

Punie, Y. & Cabrera M. (2005). The Future of ICT and Learning in the Knowledge Society. Report on a *Joint Dg Jrc-Dg Eac Workshop Seville*, 20-21 October.

Robinson, S. (2006). The mission of the j-blog: Recapturing journalistic authority online. *Journalism*, *7(1)*, 65-83.

Rosenbloom, A. (2004). The Blogosphere. *Commun ACM*, *47(12)*, 31-33.

Regan T. (2003). Weblogs threaten and inform traditional journalism. *Nieman Reports* (Fall), *57(3)*, 68-70.

Sigala M. (2008). Developing and implementing an eCRM 2.0 strategy: usage and readiness of Greek tourism firms. In O'Connor,P. Höpken, W. and Gretzel, U. (Eds) Information and Communication Technologies in Tourism 2008, (pp. 463-474.) Springer Verlag: Wien.

Sroca, N. (2007). Understanding the political influence of blogs. Institute for Politics, Democracy, & the Internet, Washington, D.C. www.ipdi.org/UploadedFiles/ PoliticalInfluenceofBlogs.pdf.

Su, N. M., Wang, Y. & Mark G. (2005). Politics as Usual in the Blogosphere. Proceedings of the 4th *International Workshop on Social Intelligence Design.*

Technorati, (2008). State of the blogosphere 2008. http://technorati.com/blogging/state-of-the-blogosphere/who-are-the-bloggers/.

Thompson, G. (2003). Weblogs, warblogs, the public sphere, and bubbles. Transactions, *7(2)*, http://transformations.cqu.edu.au /journal /issue_07 /paper_02.shtml.

Trammell, K. & Keshelashvili, A. (2005). Examining the new influencers: A self-presentation study of A-list Blogs. *Journalism and Mass-Communication Quarterly*, *82(4)*, 968-982.

Trammell, K., Williams, A., Postelnicu, M. & Landreville, K. (2006). Evolution of Online Campaigning: Increasing Interactivity in Candidate Web Sites and Blogs Through Text and Technical Features. *Mass Communication and Society*, *9(1)*, 21-44.

Walker, J. (2007). Weblog, Definition for the Routledge Encyclopedia of Narrative Theory.

Vrana, V. & Zafiropoulos, K. (2009). A social network analysis of travel blogs. In Proceedings of MCIS 2009. Information Society Research, Education, Policy and Practice in the Mediterranean Region. Athens, *September*, 25-27.

Wall, M. (2004). Blogs as Black Market Journalism: A New Paradigm for News. Berglund Center for Internet Studies, http:// bcis.pacificu.edu/journal /2004/ 02/ wall.php.

Wallsten, K. (2005). Political Blogs and the Bloggers Who Blog Them: Is the Political Blogosphere and Echo Chamber? Paper Presented at the American Political Science Association Annual Meeting Washington, D.C. September 1-4, 2005. http://www.journalism.wisc. edu/ blog-club/Site/Wallsten.pdf.

Williams, J. B. & Jacobs, J. (2004). Exploring the use of blogs as learning spaces in the higher education sector. *Australasian Journal of Educational Technology*, *20*, 232-247.

Zafiropoulos, K. & Vrana, V. (2008). A Social Networking Exploration of Political Blogging in Greece. Official Proceedings of 1st World Summit on the Knowledge Society, In: M. D., Lytras, J. M., Carroll, E. Damiani, & R. D. Tennyson, (Eds). Emerging Technologies and Information Systems for the Knowledge Society First World Summit, WSKS 2008, Athens, Greece, September 24-26, 2008. *Lecture Notes in Computer Science*, LNCS/LNAI 5288 (573-582), Springer Verlag.

In: Internet Issues: Blogging, the Digital Divide and Digital... ISBN: 978-1-60741-824-5
Editor: Cassie M. Evans, pp. 53-70

Chapter 3

UNDERSTANDING THE ACADEMIC BLOGOSPHERE: A LOOK INTO THE GENERIC TANGLE

María José Luzón
Universidad de Zaragoza, Saragossa, Spain

ABSTRACT

Blogging is becoming increasingly popular among scholars, who see it as a helpful tool to promote their research to a broad audience and to engage in discussion about discipline-specific topics and more general (sometimes ideology-laden) issues. However, although academic weblogs share many technological and structural features, they are not a homogeneous whole, since they are written by very different types of bloggers, who use this tool for various purposes. "Academic blogging" is an umbrella term that covers a wide range of hybrid genres, customised to achieve the blogger's intended purpose and to meet the expectations of the blogging community interacting through a specific blog. Therefore, we can only discuss scholars' blogging practices if we first specify clearly the contexts of use and the social and situational factors that condition weblog discourse. The aim of this chapter is to provide an overview of academic blogs by examining the complexity of the academic blogosphere and by looking into the different types of blogs inhabiting this sphere. I draw on the faceted classification scheme for CMC proposed by Herring (2007) to analyse a sample of 10 academic blogs and account for the similarities and differences between them. This analysis will help to better comprehend academic blogs and will throw light on the factors that influence performance and communicative practices in academic blogs.

1. INTRODUCTION

Weblogs (or blogs) are online genres which share features with standard webpages and with asynchronous computer-mediated communication (Herring et al., 2004). Thanks to their technical capabilities (i.e. hyperlinks, comments, trackbacks, RSS, or blogrolls), weblogs offer researchers a platform to record and organise their thinking and facilitate interaction, thus making it easier for scholars to exchange information and engage in discussion about any discipline-specific topic (Bar-Ilan, 2005; Efimova and de Moor, 2005; Ewins, 2005; Mortensen and Walker, 2002; Luzón, 2007; Paulus and Dennen, 2005). Although we identify clearly a weblog through a series of technical features, characterising the blog genre is a difficult task. Herring, Scheidt et al. (2005, p. 163) propose that a blog is a "*hybrid* of existing genres, rendered unique by the combination of features of source genres they adapt, and by their distinctive technical affordances" (emphasis in original). They also predict that "the purposes to which weblog software will be put and the conventions that will arise around them will become so diverse in the future that it will no longer be meaningful to speak of weblogs as a single genre": weblogs will become "a socio-technical format, whose convenience and general utility support a variety of uses" (Herring, Scheidt et al., 2005, p. 167). I consider that this time has already come, a view that seems to be shared by Maingueneau (2008) when he characterises blogs as hypergenres. He uses the term "hypergenre" to refer to frames that impose very poor constraints (e.g. a dialogue, a letter): they frame a wide range of texts.

Although the term "academic blog" (or science blog) restricts the scope, it is still a very broad term that includes several different genres (Walker, 2006), such as political blogs, pure research blogs, used to record and share research and ideas, or blogs about academic life. The wide variety of academic weblogs is due to the fact that they are written by very different types of bloggers (e.g. tenured professors, journal editors, PhD students, research groups), who use this tool for various purposes. Therefore, we can only discuss scholars' blogging practices if we first specify clearly the contexts of use and the social and situational factors that condition weblog discourse (Bruns and Jacobs, 2006; Schmidt, 2007). Understanding the academic blogosphere requires an analytical framework which can account both for similarities and differences among academic blogs.

The aim of this chapter is to provide an overview of academic blogs by examining the complexity of the academic blogosphere and by looking into the different types of blogs inhabiting this sphere. I draw on the faceted classification scheme for CMC proposed by Herring (2007) to analyse a sample of 10 academic blogs and account for the similarities and differences between them. This analysis will help to better comprehend academic blogs and will throw light on the factors that influence performance and communicative practices in academic blogs.

2. What Are Academic Blogs?

The amazing growth of scientific/academic blogging is attested by the increasing number of institutions, research groups and individual academics who use blogs to report and discuss their research. Authors writing on research blogging often use the terms "science weblog" and "academic blog" to refer to the same concept (although with some nuances), so I will not make any distinction between these two terms, science being understood here as a broad term including aspects related to research, academia, etc. in any discipline. In his blog *Science and Politics* Zivkovic (2006) points out that science blogs are defined by bloggers themselves in two ways: "by topic-matter and by authority of the author. Thus, a science blog is one that always, often, or at least sometimes covers science as a topic. Or, a science blog is one written by a person with some expertise in science, e.g., a practicing scientist, a student, or perhaps a science journalist." Torres-Salinas and Cabezas-Clavijo (2008) examine the implications of adopting each of these perspectives. Accepting that any blog that to some degree includes postings dealing with science is a science blog involves including a high number of blogs[1], with very different topics, perspectives- going from technology to pseudoscience (Bonetta, 2007)- and quality, some of which have absolutely no interest for the scientific community. Therefore, Torres-Salinas and Cabezas-Clavijo propose adopting both criteria to consider a blog as academic/ scientific: they should be published by competent members of the scientific and academic community (e.g. researchers, professors, PhD students, even scientific journalists) and content should be exclusively limited to scientific aspects, with no incursion in personal aspects. However, although scientific blogs should certainly focus on scientific topics, very rarely will we find a scientific/ academic blog where no post goes away from science into other spheres.

Coleman (2005) considers that academic blogs have the following features: open and free access; written for professors, students and/or practitioners; including comment or other discussion link mechanism; archive; a core value, belief or set of intellectual concepts on which posts are focused; current and updated weekly or more frequently; responsive to relevant global news; accurate posts; bloggers with authority in a clearly defined area; a community of readers. However, again, many of these criteria are frequent but not essential features of academic blogs. For instance, regarding the audience, many blogs are not only intended for experts, but also try to get to a lay audience. The author of *In the Pipeline* (http://www.corante.com/pipeline/), for instance, points out that he writes posts for a broad readership, from chemists, to investors, to the lay public (cited in Bonetta, 2007). Similarly, although the comment feature is included in most academic blogs, there are quite a few that do not allow comments.

The difficulty to delimit academic blogs derives also from the fact that even if we can make some distinction in terms of content, such as blogs for political exchanges, pure research blogs, or blogs about academic life (Walker, 2006), most academic blogs are in fact hybrid genres, with different types of entries (Davies and Merchant, 2007). Zivkovic (2006) provides a list of the kind of posts that can be found in academic blogs. Among these he includes: occasional personal posts (about family, hobbies, etc); daily news and events, be it related to science, politics or popular culture; humour; daily life in the laboratory or the field (mainly in blogs written by students and postdocs); book reviews; comments on new scientific findings as reported by the mainstream media; politics of science; science and religion; science education; comment of recent papers to teach the audience about a broader area of scientific research (what Zivkovic calls "placing science news into context"); science reviews and tutorials, etc. In spite of this variety of topics, both Zivkovic (2006) and Bonetta

[1] In 2007 Bonetta cited Bora Zivkovic's (http://scienceblogs.com/clock) estimation of the number of science blogs as 1,000 to 1,200, considering science blogs those "written by graduate students, postdocs and young faculty, a few by undergraduates and tenured faculty, several by science teachers, and just a few by professional journalists." More recently the post "How many (science) blogs are there" (2008) in the blog *Living the scientific life* (http://scienceblogs.com/grrlscientist/2008/09/how_many_blogs_are_there.php) sparked an interesting discussion, with participants agreeing that it is impossible to calculate the number of science blogs, because first it is impossible to delimit what a scientific blog is. Participants considered that the blogger being a scientist does not automatically imply that he/she writes a science blog. As a way of example, one of the participants mentions *Pharyngula* (http://scienceblogs.com/pharyngula/) (occurring as the most popular

(2007) point out that most scientist bloggers do not talk about their unpublished research or that of their colleagues. However, Bonetta (2007) mentions a blog that is an exception to this rule: the blog *Useful Chemistry* (http://usefulchem.blogspot.com), by Jean-Claude Bradley and his students at Drexel University, presents and analyses results, with links to the raw data on the wiki, with observations and conclusions. As Bonetta puts it, they include "every detail a scientist would include in a lab notebook except that the information is available on the Web for everyone to see and comment on" (2007, p. 445). This may be the precedent for another academic online genre, where the ideas of Science 2.0. result in a more open sharing of research.

3. Uses of Academic Blogs: Why Do Scholars Blog?

In 2003 Glenn asked the following question "Is this a revolution in academic discourse, or is it CB radio?" in *The Chronicle of Higher Education* (Glenn 2003) to introduce a column on the reasons why scholars blog. The question is still unanswered, or rather, it has multiple answers, since not all scholar bloggers have the same motivations.

A large number of papers and blog posts have discussed the uses of weblogs in research and academia (Bar-Ilan, 2005; Davies and Merchant, 2007; Efimova, 2004; Efimova and de Moor, 2005; Glenn, 2003; Halavais, 2006; Luzón, 2006, 2007; Mortensen and Walker, 2002; Paquet, 2002; Torres-Salinas and Cabezas-Clavijo, 2008; Walker, 2006). These researchers have examined the affordances provided by blog technologies and present blogs as facilitating scientific enquiry in two ways: (i) they help to disseminate information and to access and manage content, through features such as links, archives, RSS, searchable databases, post categories; and (ii) they are tools for collaboration and interaction. As a tool for content management and information spreading, weblogs help to manage personal knowledge by allowing bloggers to keep notes on their reading and thinking. Additionally, they provide pointers to and evaluation of relevant resources. An interesting initiative in this regard is that of *BPR3.org* (Bloggers for Peer Reviewed Research Reporting), a venture initiated by Dave Munger of *Cognitive Daily* to encourage the dissemination of valuable research. They provide an icon to identify posts with peer-reviewed work, which may help readers to distinguish these posts from news, politics, personal issues, and so on. As tools for discussion and collaboration, blogs have several uses: supporting community forming, i.e. helping to create a virtual community of people interested in the same topic; facilitating connections between researchers and developing, maintaining and activating connections with others; supporting the discussion of topics or theories in which several researchers are interested, thus contributing to developing or articulating an idea and to reinforcing the links between the members of the virtual community. Batts, Anthics and Smith (2008) illustrate how blogs may have a significant impact on traditional academia by reporting on a case where a science blogger became coauthor on a paper in *Plant Cell*. In 2005, Reed Cartwright, postdoctoral fellow and blogger at *De Rerum Natura* (http://dererumnatura.us/), used his blog to disagree with the conclusions of a paper in *Nature* and put forward an alternative interpretation of the findings. Luca Comai at the University of California had developed a similar interpretation

science blog in many top lists), which has a very small percentage of science posts and, therefore, might not qualify as science blog.

and when he was in the process of publishing it in the journal *Plant Cell*, he found out that Cartwright had already posted that idea in his blog, and offered him to be a coauthor of the *Plant Cell* paper.

In addition to facilitating scientific enquiry, blogs also increase the bloggers' (or the institution's) visibility. Previous research has revealed that most academic blogs have a promotional purpose, since many entries are used to publicise research by an academic or research group, to announce activities, events and relevant news related to an institution or group, or to make available to the public the researchers' work, usually with the purpose of getting feedback (Luzón, 2006; Torres-Salinas and Cabezas-Clavijo, 2008). Most academics use blogs for self-presentation, in order to increase their visibility and develop respect and reputation (Davies and Merchant, 2007; Luzón, 2006). In a live online discussion, scholar blogger Ed Voloknov pointed out that

> Blogging is also a mechanism for scholars to promote their scholarship to a broader audience, both scholarly and non-scholarly. So even if scholars won't get faculty credit for blogging as such, they'll get more satisfaction because their ideas are being more widely spread -- and they may eventually get a better academic reputation as a result, precisely because more people will be noticing their ideas. (quoted in Coleman, 2005, p. 5).

There is another motivation that has been devoted little attention in the literature on academic blogging, but helps to account for many entries in this kind of blogs: the ethical and social motivation. In the online discussion mentioned above, Voloknov also noted that "blogging will eventually come to be valued the way that op-eds or media appearances are valued -- at my institution, we'd call it part of the person's service to the community, rather than part of the person's scholarship" (quoted in Coleman, 2005, p. 5). Batts et al. (2008) also mention the opportunity for bloggers to present research in an accessible format and thus "provide a unique educational bridge between academia and the public". Similarly, among the reasons for blogging that Brown (2009), a post-doctoral fellow writing the blog *Biochemical soul* (http://biochemicalsoul.com/), mentions are the following: "Hone your ability to distill complex research into understandable terms" and, interestingly, "Be a part of 'the good fight'", i.e. engaging in pro-science activism against those who want to impose ignorance on this and future generations.

Academic blogs are also a way of "personal academic disclosure", that is, not a way to disclose aspects of the bloggers' personal life (which many scholar bloggers also do), but of showcasing their academic persona, i.e. a way to show that research is not only their job, but also their passion. This is clearly revealed in some of the reasons that Brown gives for blogging in the post "Science Blogging: The Future of Science Communication & Why You Should be a Part of it" (2009): "Share your passion for science", "Showcase your dedication to science and your interest in outreach". In his view, the blog is a tangible accomplishment to show the things the blogger is interested in as a person, to show how much he/she loves science, and to reach the lay public. Even when he mentions being part of a community as a reason to blog, Brown focuses on the emotional component, rather than on the career benefits of being part of the community:

> I've found few communities quite like that of the science bloggers. We are nerds and proud of it. We care deeply about all aspects of science, with particular interest in bringing it

to the people. From the first day of becoming a part of this community, you can rest easy knowing that you have just entered a virtual world *full* of people that are like you in many respects.

4. Types of Academic Blogs

Although, as pointed out above, we could use topic and authorship as criteria to define academic blogs, those blogs cover a wide variety of topics and are written by different kinds of authors. Using content as a criterion, Walker (2006) makes a distinction between three types of blogs, which she refers to as "public intellectuals", "research blogs", and "blogs about academic life". She uses the term "public intellectuals" to refer to blogs that are used as platforms for political debate, with various underlying perspectives. "Pure research blogs" are used to record and share research and ideas, and "blogs about academic life" are those where the scholar discusses general aspects of academic life and research (e.g. the process of earning a PhD or tenure and so on). This is a good distinction for working purposes. However, as already noted above, most academic blogs tend to be hybrid, with different kinds of posts (i.e. a blog with a predominant number of posts discussing research, may include, and usually includes, some post discussing political issues, more general academic aspects, or even personal matters).

Here I use authorship as the criterion to differentiate among types of academic blogs. I will briefly discuss the following: institutional blogs, blogs by research groups, group blogs, blogs by journal editors, blogs by professors, PhD blogs.

4.1. Institutional Blogs

Batts et al. (2008) point to different ways in which institutions seek to expand their presence in the blogosphere. First, they may provide a hub for intra-institutional blogs or an aggregate feed of the faculty and student blogs at the institution. This is the case of the feed at the *Oxford Internet Institute* (OII) (http://people.oii.ox.ac.uk/). Batts et al. stress the benefit of this set up both for the institution, "which gets free publicity for its researchers' work", and for the academic bloggers, "who have a built-in readership funneled straight from the institution's Web page". Institutions may also have their own blog, with posts by faculty members, as in the case of *Rudd Sound Bites*, the blog of Yale University's Rudd Center for Food Policy and Obesity Research (http://www.ruddsoundbites.typepad.com/).

4.2. Blogs by Research Groups

Research group blogs are those written by members of a research group, usually faculty and staff within the same institution. After analysing a corpus of research group blogs, Luzón (2006) found that the entries in these blogs have the following main purposes: (i) publicising the group and the group's research; (ii) making available to the public the members' work, usually with the purpose of getting feedback; (iii) communicating with other members of the group: organising and coordinating the group's work and managing knowledge exchange within the group; (iv) creating a sense of community, becoming more visible in this virtual community and seeking collaboration easily; (v) social function: providing socioaffective support and reinforcing social links (by sharing personal anecdotes, congratulating on personal and academic issues, etc.).

4.3. Group Blogs

These are blogs with several contributing members, usually from different institutions, who are experts in a field. This discipline membership is the main relation between the participating members. Some examples are *Cosmic Variance* (http://blogs.discovermagazine.com/cosmicvariance/) for physics, *Real Climate* (http://realclimate.org/) for climate science, or *Savage Minds* (http://savageminds.org/) for Anthropology. Some blogs written by more than one blogger would not be considered here group blogs, because the individual's belonging to the group is based on offline relationships. A case in point is, for instance, that of *Cognitive Daily* (http://scienceblogs.com/cognitivedaily/), whose writers are Greta Munger (Professor of Psychology at Davidson College) and her husband, Dave Munger, co-founder and president of ResearchBlogging.org.

4.4. Journal Blogs

Here I include both blogs set within a journal and blogs by journal editors. There is a high number of scientific journal blogs, e.g. *The Scientist*'s blog (http://www.the-scientist.com/blog/), described by the creators as "the place where professionals from all sides of the life sciences get together to aggregate, share and discuss our experiences", or the *Scientific American*'s blog (http://www.scientificamerican.com/blog/60-second-science/), described in the *Scientific American* website as "providing easily digestible explanations of breaking scientific news and describing how developments can change the environment, our technology and your health". The news blog in *Scientific American* is written by their reporters and editors and they state that they sometimes include comical, quirky or bizarre stories, that way showing "that just because it's science doesn't mean it can't be fun". The blog *Blog.bioethics.net* is written by the editors of *The Americal Journal of Bioethics*. Their reason for writing a blog is that "the blog format allows the editors to tap an incredible stream of information that relates to bioethics as it flows through our offices": the blog fills a gap, attending to people's demand for a place which covered all the news, discussed them or put them in context, something that couldn't be done in the journal's website.

4.5. Blogs by Professors

Focusing on legal blogging, Schilt (2007) observes that most bloggers are not young academics but mid-career tenured professors. The benefits of blogging for these academics are, among others, wider distribution and readership than traditional scholarship, immediate feedback and discussion and increased recognition in their field. However, there is a high number of researchers from other categories (e.g. non-tenured professors, research fellows) who write academic blogs. Luzón (2008) analysed blogs by three different groups of people within academia (PhD students, non-tenured professors and tenured professors) and found that different groups of academics according to status may engage in blogging for different purposes and may use blogs for setting up different types of relations with other members of the disciplinary community

4.6. Blogs by New Scholars and PhD Students

Although Schilt (2007) points out that younger scholars might consider that blogging is not worth the time taken away from traditional scholarship, there is an impressive number of blogs by young scholars and PhD students in almost any discipline. Almjeld (2006) remarks that blogs are quite helpful for new scholars "as storehouses of scholarly work, future writing topics, and colleague feedback". Among the benefits of blogging for new scholars he mentions the following: the ability to organise previous papers and ideas for later retrieval, the ability to test out new knowledge and ideas in a safe environment, the pressure to further elaborate and explore ideas and connections, the ability to get genuine comments and useful feedback, and the ability to show one's skills and interests and thus find one's place in the field. In their analysis of the social dimension of blogs maintained by university and graduate students, Nardi, Schiano and Gumbrecht (2004) found the following motivations for blogging: update others on activities and whereabouts, express opinions to influence others, ask for others' opinions and feedback, 'think by writing', and release emotional tension. Luzón (2008) points out that even within PhD blogs there is variety: although most of them seem to be used as a diary to record PhD progress, personal aspects and worries about academic life, some of them are more "professional" and are intended for discussion of disciplinary topics.

5. Methodology

5.1. Method

In order to shed some light on the complexity of the academic blogosphere I use the faceted classification scheme for CMD (computer-mediated discourse) proposed by Herring (2007) to analyse a sample of 10 academic blogs and account for the similarities and differences between them. Drawing on the assumption that CMD is subject to two basic types of influence, i.e. medium (technological) and situation (social), Herring (2007) proposes a scheme including 10 medium facets (categories) and eight situation facets, with several possible realizations (terms) for each. Both the lists of medium and situation factors are considered as open-ended. The medium categories, which describe technological features of CMC systems, include: synchronicity (values: synchronous/ asynchronous), message transmission (values: 1-way/2-way), persistence of transcript (scalar values: degree of persistence of text, from low to high), size of message buffer (i.e. the number of characters the system allows in a single message), channels of communcation (how many and what kinds), anonymous messaging, private messaging, filtering, quoting, message format (i.e. order in which messages appear, how they are visually presented, etc). As for the social and situational factors that are considered by Herring to condition variation in CMD, they are presented in Table 1 (adapted from Herring, 2007).

5.2. Analysed Sample

In order to carry out the analysis I selected 10 academic blogs written in English, 2 from each of the following categories: institutional blogs, blogs by research groups, group blogs, blogs by professors, blogs by PhD students (see appendix I). I selected 15 messages from each blog, beginning on June 1st, in order to characterise them in terms of the facets included in the scheme described above, and thus, compare blogs within and across categories.

I have mainly compared them in terms of situation features, since, as Herring points out (2007), when she herself compares two blogs, blogs share many medium features, which could be considered "definitional characteristics of the blog genre", e.g. asynchronicity; 1-way message transmission; persistence of messages in archives linked from the sidebar of the blog; Web-based delivery; the display of blog entries in reverse chronological sequence with a "comment" option below each entry (Herring, 2007). There are admittedly other medium features in which blogs may differ, e.g. although the channel of communication is mostly text, some blogs include a high number of graphics and videos, or podcasts of the postings; some blogs have more sophisticated interfaces than others, providing users with more options regarding message format.

Table 1. Situation factors that condition variation in computer-mediated discourse

S1	Participation structure	One-to-one, one-to-many, many-to-many Public/ private Degree of anonymity, pseudonymity Group size: number of active participants Amount, rate and balance of participation
S2	Participant characteristics	Demographics: gender, age, occupation, etc Role/status: in "real life"; of online personae Pre-existing sociocultural knowledge and interactional norms Attitudes, beliefs, ideologies and motivations
S3	Purpose	Of group, e.g. professional, social, role-playing Goal of interaction, e.g. get information, negotiate consensus, have fun
S4	Topic or theme	Of group, e.g. politics, science fiction Of exchanges, e.g. the economic crisis, vacation plans
S5	Tone	Serious/ playful Formal/ casual Contentious/ friendly Cooperative/ sarcastic, etc.
S6	Activity	e.g. debate, information exchange, joking, flirtation
S7	Norms	Of organisation Of social appropriateness (i.e. "behavioural standards of the group") Of language (i.e. "linguistic conventions particular to a group of users")
S8	Code	Language, language variety

6. RESULTS

6.1. Institutional Blogs

The two institutional blogs analysed are *Rudd Sound Bites (RSB),* the blog of the Rudd Center for Food Policy and Obesity at Yale University (http://www.ruddsoundbites.typepad.com/) and *Beck Institute Blog (BIB)* (http://www.beckinstituteblog.org/), the blog of the Beck Institute for Cognitive Therapy and Research. The two blogs are open to the public and in both several authors have posting rights. Another common feature is the low frequency in the postings: the 15 posts take a span period of three and a half months (from May 11th to January 30th, 2009) in the *Beck Institute Blog* and of two months (from June 1st to March 30th, 2009) in *Rudd Sound Bites*. There are also similarities in the participant characteristics: variety of gender and age, different roles, but all of them within the academic sphere, attitudes and beliefs shared by all the authors in the blog.

However, there are also differences. The number of active participants is different: while in *RSB* there is a high number of active participants writing entries, an author index with 27 people, and the author of any post is identified, in the case of the *BIB* we don't really know the number of staff writing posts and there is no information about the author of individual posts. Additionally, while all the posts in the *RSB* blog got comments (between 3 and 19), none of the posts in the *BIB* got comments.

The purpose is made explicit in the "about page" of both blogs. The purpose of the *Rudd Sound Bites Blog* is "to encourage a free exchange of ideas about obesity and food policy in our modern world". And that of the *Beck Institute Blog* is to provide a resource on Cognitive Therapy for everyone (professionals, students and the general public), to answer reader's questions about what Cognitive Therapy is and how it works, to provide a platform to share their experiences and to inform about events at the Beck Institute. The activities are highly related to the purposes. The main activity in the *Rudd Sound Bites Blog* is discussion of news, laws, advertising, etc. related to obesity and food policy. The main activities in the *Beck Institute Blog* are author's discussion of studies on Cognitive Therapy, announcement of and report on events at the Institute (e.g. training workshops), link to and comments on Cognitive Therapy in the news.

Regarding norms, the *RSB* provides explicit statement of how interaction should develop: "For us, the true mark of success is an active community of readers who critique our arguments and feel free to respectfully disagree (…) Therefore, we encourage not just discussion, but disagreement. In return, we ask that our participants express their views with respect and courtesy of others". As for the *BIB*, since it is not intended as a forum for exchange of ideas and discussion, there is no reference to norms of interaction.

There is a difference in tone. While the posts in *BIB* are quite formal in tone, those in *RSB* are much more informal. The following example illustrates the tone of posts in *RSB*:

> (1) Jones-ing for a less guilt-inducing sweet snack the other day, I mindlessly grabbed the Klondike "100 Calorie" ice cream bars on sale without paying attention to the nutrition labeling on the back. The pictures of ice cream enrobed in chocolate with the low-cal promise were good enough for me (see Exhibit A). Later that night, I dashed to the freezer to enjoy a treat (*Rudd Sound Bites*).

6.2. Research Group Blogs

The two research group blogs analysed are the blog by the *Princeton S* Network Systems* (SNS) group (http://sns.cs.princeton.edu/blog/) and *Nmrg* (http://spooner.beds.ac.uk/nmrg/), the weblog of the new media research group of the Research Institute for Media, Art and Design at the University of Bedfordshire, UK. Although in most academic blogs the dominant mode in postings is text, in *Nmrg*, it being the blog of a group carrying our research on new media, video is the predominant medium (i.e. 11 out of the 15 postings only have a video file, with no added commentary text).

The two samples have several situation variables in common: one-to-many participation structure, open to the public, open disclosure of the author's identity (no anonymity). Both blogs also share the low frequency in the postings: the 15 posts in *Nmrg* take a span period of 6 months (from March 12th, 2009 to October 13th, 2008). As for the *Princeton S* Network Systems* blog, there are only 12 postings in the blog, in the period from June, 11th, to March 12th, 2009. There are also similarities in the participant characteristics: variety of gender and age, all the authors within the academic sphere (although with different status). Besides, all the authors in each of the blogs belong to the same research group and, presumably, share many attitudes, beliefs, and disciplinary knowledge. In both cases, there is an imbalanced participation among the blog authors. *Nmrg* is presented as "the group blog for the new media research group in RIMAD at the University of Bedfordshire, UK". However, although there are 11 potential authors (i.e. 11 people in the group), 14 of the 15 posts in the sample are written by the same author. Similarly, the Princeton S* Network Systems group has 8 members, but 8 out of the 12 posts are by the group leader. Another common feature is that although comments are allowed, they are non-existent (*Nrmg*) or very rare. In the *Princeton S* Network Systems* blog only three blogs got comments (2, 3, and 10 comments).

The purpose of the posts is also similar in both samples. They are used to exchange content that can be of interest to the other members of the group or to a broader audience in the discipline, to announce and report events, specially those taking place in their workplace (e.g. conferences, fellowship programs) and to publicise their work and that of their institution (see example 2).

> (2) We have just received some very good news here at the *Research Institute for Media, Art and Design*. In the recent research assessment exercise half our research was rated world leading or internationally excellent and 95% was judged as being of international significance (*Nmrg*).

As for the content, it is highly discipline-focused in both cases. *Nmrg* focuses on issues related to new media. As pointed out about, the blog includes many videos, most of them from *YouTube*. Some of them are creative videos, offered in the blog as examples of art and design, and some other are videos of presentations on the topic. The tone is also similar in both blogs: formal, but friendly and cooperative.

6.3. Group Blogs

The two group blogs analysed are *Cosmic Variance* (http://blogs.discovermagazine.com/cosmicvariance/) and *Real Climate* (http://realclimate.org/). In both of them, there are links to the comments and to trackbacks, which are called "blog reactions" in *Real Climate*. Posts in blogs are not often commented or reacted to in other blogs, but in *Real Climate* many posts have blog reactions. In addition, some of the blogs in *Real Climate* have a link to their translation into another language. These medium features show that the blog is intended as a global discussion forum, and that there is a high interest in integrating and giving access to all the perspectives from which an issue can be discussed.

The two samples have the following situation variables in common: open to the public, open disclosure of the author's identity (no anonymity). All posts in both blogs are signed by the author(s), except "group posts" in *Real Climate*, which are collective efforts from the whole team. Other common features are imbalanced participation between authors and readers, with authors writing longer posts than readers, high posting frequency, and high number of comments, when compared with other academic blogs: the comments to the posts in *Cosmic Variance* ranged from 5 to 59, and in *Real Climate* from 42 to 1154.

There are also many similarities in the participant characteristics: multiple authors (7 in *Cosmic Variance*, and 11 in *Real Climate*), authors are both male and female, they are all reputed scientists working at universities or research centres, all the authors in the blog share similar attitudes, beliefs and disciplinary knowledge. *Cosmic Variance* is described as "a group blog by people who, coincidentally or not, all happen to be physicists and astrophysicists". Although in both blogs there are some authors that write more posts, author participation tends to be quite balanced.

Regarding purpose, *Real Climate* is presented as a commentary site on climate science by working climate scientists for the interested public and journalists, intended "to provide a quick response to developing stories and provide the context sometimes missing in mainstream commentary". *Cosmic Variance* has a similar discipline-related purpose. The activities in both blogs are very similar: debate, comment on news, tutorials on aspects related to climate (for students), exchange of information (e.g. presenting papers) in order to trigger off discussion on results.

Regarding topic, there is a difference. The bloggers in *Cosmic Variance* write about science, mainly physics and astrophysics, but also about whatever they find interesting, in their words: "arts, politics, culture, technology, academia, and miscellaneous trivia". Nevertheless, in contrast with other types of academic blogs, there are no posts about the blogger's personal work progress. In *Real Climate* it is explicitly stated that the discussion "is restricted to scientific topics and will not get involved in any political or economic implications of the science". However, as a matter of fact, although all their posts deal with climate issues, it is not always true that they do not deal with political or economic implications of science.

As for the norms of interaction, both of them are moderated forums, and foster discussion with disagreement. The authors of *Cosmic Variance* state: "We love comments and aim to cultivate a lively and enjoyable space for discussion. To this end, we will not hesitate to delete comments or ban commenters who are excessively impolite or who otherwise derail the discussions. *Disagreement* with anything we may say is welcome, so long as it is civil and constructive. We're all about light, not heat." Discussions in *Real Climate* do get relatively heated sometimes, but a moderate degree of heat seems to be acceptable. In both samples the tone is relatively serious and formal, but friendly and highly cooperative.

6.4. Blogs by PhD Students

The two PhD blogs analysed are *Life of a PhD student* (http://lifeandtimesofaliteraturephdstudent.blogspot.com/) and *Developing Intelligence (DI)* (http://scienceblogs.com/developingintelligence/). The two samples have the following situation variables in common: one-to-many participation structure, open to the public, single author. The main similarity is that the bloggers have the same status within the academic community and are engaged in PhD writing.

There are, however, many differences. While *Life of a PhD student* is an anonymous blog written by a female student, the blogger at *DI* is a male student who gives a great deal of information about himself (his name, his academic life, his personal life). The only information that the author of *Life of a PhD student* provides about herself is that she is in the writing up year of her PhD in Renaissance literature. She does not either mention other names that may help identify her: she uses the noun "Boy" for her boyfriend and "Supervisor" for her supervisor.

The purpose is also different. *Life of a PhD student is a* personal journal blog of the author's daily life, who states "I look to blogging to keep me sane". The purpose of *Developing Intelligence* is to keep a record of the author's readings and ideas. He states: "I started a blog so that I can remember my thoughts about new papers". The difference in purpose is reflected in activity and topic. The blog *Life of a PhD student* has a sub-title: "the random ramblings of a graduate student struggling to succeed" and both the title and the subtitle give a clue as to the nature of this blog. It is really a diary weblog, where the author talks about her daily life (e.g. what she did during Bank Holiday), her hobbies, her personal relationships, her progress in the PhD work (e.g. 3), how she is feeling, etc.

> (3) At present I am completing a full draft of the first chapter of my thesis. This chapter was already drafted, so I suppose it is not entirely the 'first full draft', more like the second, or third, or twenty-third. But, to me, it's the first truly functional draft in that I am mostly satisfied that it now makes sense, is well ordered, and has the correct information in it (*Life of a PhD student).*

In another post she informs about the acceptance for publication of one of her papers, but the purpose is not to make it available to the public (there is no link) or to show her competence, but to share her happiness. The tags for two of the posts in the sample give an idea of the content of the blog: chapter one, isolation, loneliness, PhD, PhD depression, supervisor, thesis; deer, friends, garden, house, jays, lunch, peace, Scottish Bank Holiday.

As for *DI,* it is used by the blogger to present, evaluate and discuss research in the field of cognitive neuroscience, comparative psychology and artificial intelligence. All the posts are discipline-related and all the posts in the sample are identified with the symbol 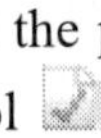 to show that peer-reviewed research is discussed.

The difference in purpose is also reflected in the number and type of comments. Although the comment feature is enabled in both blogs, only two posts in *Life of a PhD student* have comments (one comment each post), probably from friends. One of them congratulates her on the publication ("I'm so pleased for you! And a little bit jealous…") and the other is used to show sympathy with the blogger's feelings. In *DI* most blogs got comments (from 1 to 18) from readers, who used them to discuss the research presented in the posts. The tone is also different. While in *The Life of a PhD student* it is casual, friendly and intimate, in *DI* it is very formal, similar to that of research paper writing.

These two blogs exemplify two types of academic PhD blogs. As a matter of fact, the inclusion of *The Life of a PhD student* within the category of academic blogs is highly questionable: although it is written by a PhD student, none of the posts is used to discuss discipline-related issues. There are, however, a lot of blogs similar to *The Life of a PhD student* (i.e. blogs where student writers detail their personal and work life), as can be seen in *Blogs about: PhD* (http://en.wordpress.com/tag/phd/). Two good examples, also written by female students who do not reveal their identity, are *YoungFemaleScientist* (http://youngfemalescientist.blogspot.com/) and *Disgruntled Julie; a PhD in progress* (http://ethidiumbromide.blogspot.com/). The only information that the authors write about themselves is the following: "I'm a postdoc in the biomedical sciences. I'm a member of several professional societies" (*Young FemaleScientist*) and "I'm Disgruntled Julie, a Ph.D. student in oncology, doing biochemical and biophysical research. This is where I write about my experiences as a graduate student, with some crafting and recipes (and the adventures of being married to a doctor; soon to be long-distance!) thrown in on the side." (*Disgruntled Julie*). Other PhD students write blogs more similar to *DI*. This is the case of *PhD in History* (http://phdinhistory.blogspot.com/), a blog whose author provides a lot of information about himself (name, university, work as a part-time teacher). Like *DI*, the blog is discipline-focused, with no personal issues, and virtually all postings get comments from readers.

6.5. Blogs by Professors

The two blogs by professors analysed are *Aetiology* (http://scienceblogs.com/aetiology/) and *Culture cat: Rhetoric and Feminism* (http://culturecat.net/). Both have a one-to-many participation structure and are written by female assistant professors who provide a lot of information about their academic status and position and about their research fields.

However, the purpose of the blogs, the topic and the academic persona that bloggers project is not exactly the same. *Aetiology* is used as a forum for discussion and debate on a wide variety of issues pertaining to the science of biology. This is reflected in the number and type of comments that posts get (from 5 to 31 in the sample). Regarding the topic, the blogger herself states: "while much of the focus is on the epidemiology of infectious disease or general microbiology, these are fields that lend themselves to discussion of a much wider range of topics: general epidemiology, vaccines, medicine, public health, pseudoscience, and politics, just to name a handful". The tags of some of the last posts show the topics dealt with in the blog: activism, politics, public health, infectious disease, intelligent design/creationism, general epidemiology, general biology, influenza, outbreak. Just to give an example of activism, the title for the post of June 1st 2009 is "silence is the enemy". It deals with rape during wartime and the author encourages readers to help stop it: "It needs to stop, and we can help make that happen. *Talk to someone about this.*" Although the blog is not strictly restricted to discipline content, there are not, however, posts dealing with the blogger's personal life and family.

Culture Cat is much more personal and similar to a diary weblog. The blogger often writes about her baby (and posts pictures of him) or her husband, and mixes domestic issues with reflections on academic workload. There are also posts on the work being carried out in the research projects she is involved, or comments on issues related to rhetoric, composition pedagogy, etc. A good example of the nature of the blog is provided in the post "Summer projects", where she writes a list of the things she intends to do during the summer, under four categories: research (e.g. "Book proposal: get it done, send it out"), teaching (e.g. "Prepare for first-year writing course I'm teaching in the fall"), home improvement (e.g. "Clean out dining room closets"), and miscellaneous ("Take one-day pediatric CPR/first aid class"). The blog is used, therefore, to keep notes both of personal/ domestic and academic life. This is confirmed by the fact that there are few comments and most of them are personal (e.g. congratulations when she posts the picture of his child's birthday) or cooperative (e.g. offering help to solve some problem).

The tone in both blogs is also quite different. Although the tone in *Aetiology* is informal, a certain social distance is kept, and contentious statements and heated debate may occur. The tone in *Culture Cat* is more casual and friendly.

Culture cat seems to represent the personal end of a continuum in individual academic blogging, ranging from blogs where all or most posts deal within discipline issues, going through topics such as academic, politics, etc. up to personal issues. Although blogs written by individual professors deal mostly with discipline and academia related issues, most of them include references to their personal life to a larger or lesser degree.

7. Conclusion

Starting from the assumption that blogs are a socio-technical format, whose technical affordances lead to their use for a variety of purposes (Herring, Scheidt et al., 2005), and that even academic blogs are not a homogeneous whole, I have explored the complexity of the academic blogosphere by using the faceted classification scheme for CMD (computer-mediated discourse) proposed by Herring (2007). The purpose of this chapter is not to use this scheme to classify academic blogs in terms of their medium and social features, but to bring to light that academic blogging is a complex phenomenon which encompasses a great variety of discourse practices, mostly deriving from the features and motivations of the bloggers.

The comparison of the blogs in the sample suggests that the academic weblog includes several subtypes which have some features that may help to identify them. The most prominent features to distinguish between academic blogs seem to be the following: the number of author bloggers and the relations between them, the participant characteristics (i.e. gender, position or status in the academic community), the purposes and motivations for blogging, which are highly influenced by the participant characteristics, and, in turn, determine the content and activities of the blog. For instance, there are many similarities between the two samples of group blogs analysed (*Cosmic Variance* and *Real Climate*), mostly deriving from the fact that they are blogs by groups of scientists sharing attitudes and disciplinary knowledge, used for discussion of discipline issues. Other blogs, such as those by staff in an institution or those by members of a research group, tend to include posts with a promotional purpose. And many blogs by individual scholars include posts on personal issues. The analysis has also confirmed that academic blogs are often difficult to assign to a subtype, since they are hybrid artifacts, and that they tend to combine several purposes, and incorporate a variety of elements: anything that the bloggers consider useful for their purposes and that is afforded by the technology. As pointed out above, the sample that I have used is very limited, both in terms of the number of blogs and the number of posts selected for analysis. A larger scale study, with a high number of blogs from each sub-type, and a higher number of posts taken from each blog, could help to provide a more detailed characterisation of each type and to reveal those features (if any) that are defining or prominent in these types.

In spite of the differences between them, what all academic blogs share is the openness and the desire to share experiences, knowledge, ideas, and to contribute to the rise of a new, more dynamic, connected and collegial model of scholarly interaction, which helps to increase productivity through sharing. Academic blogs are the seed to this open sharing and discussion.

REFERENCES

Almjeld, J. (2006). Making Blogs Produce: Using Modern Academic Storehouses and Factories. *Computer and Composition Online*. Retrieved July 14, 2009 from http://www.bgsu.edu/cconline/almjeld/gradblogs.htm.

Bar-Ilan, J. (2005). Information hub blogs. *Journal of Information Science*, *31(4)*, 297-307.

Batts, S. A., Anthis, N. J. & Smith, T. C. (2008). Advancing Science through Conversations: Bridging the Gap between Blogs and the Academy. *PLoS Biol, 6(9)*, e240. doi:10.1371/journal.pbio.0060240.

Bonetta, L. (2007). Scientists enter the blogosphere. *Cell*, *129(3)*, 443-445.

Brown, D. D. (2009). Science Blogging: The Future of Science Communication & Why You Should be a Part of it. Retrieved July 14, 2009 from http://biochemicalsoul. com/2009/03/science-blogging-the-future-of-science-communication-why-you-should-be-a-part-of-it/.

Bruns, A. & Jacobs, J. (2006). Introduction. In A. Bruns, & J. Jacobs (Eds.), *Uses of Blogs* (pp. 1-8). New York: Peter Lang.

Coleman, A. (2005). Academic blogs. Retrieved July 12, 2009 from http://dlist.sir.arizona.edu/1311/01/ablogascfinal.pdf.

Davies, J. & Merchant, G. (2007). Looking from the inside out: academic blogging as new literacy. In C. Lankshear & M. Knobel (Eds.), *A New Literacies Sampler* (167-198). New York: Peter Lang.

Efimova, L. (2004). Discovering the Iceberg of Knowledge Work. *Proc. of the Fifth European Conference on Organisational Knowledge, Learning, and Capabilities (OKLC 2004)*. Innsbruck, April Retrieved June 30, 2009, from https://doc.telin.nl/dscgi/ds.py/Get/File-34786.

Efimova, L. & de Moor, A. (2005). Beyond personal webpublishing: An exploratory study of conversational blogging practices. In *Proceedings of the Thirty-eighth Hawaii*

International Conference on System Sciences (HICSS-38). Los Alamitos: IEEE Press. Retrieved July 11, 2009 from https://doc.telin.nl/dsweb/Get/Document-44480/

Ewins, R. (2005). Who are You? Weblogs and Academic Identity. *E-Learning*, *2(4)*, 368-377. Retrieved July 14, 2009 http://dx.doi.org/10.2304/elea.2005.2.4.368.

Glenn, D. (2003). Scholars who blog. *The Chronicle of Higher Education*. Retrieved July 14, 2009 from http://chronicle.com/free/v49/i39/39a01401.htm.

Halavais, A. (2006). Scholarly blogging: Moving towards the visible college. In: A. Bruns, & J. Jacobs, (Eds.), *Uses of Blogs* (117-126). New York: Peter Lang.

Herring, S. C., Kouper, I., Scheidt, L. A. & Wright, E. (2004). Women and children last: The discursive construction of weblogs. In: L. Gurak, S. Antonijevic, L. Johnson, C. Ratliff, & J. Reyman (Eds.), *Into the Blogosphere. Rhetoric, Community, and Culture of Weblogs*. Retrieved July 14, 2009 from http://blog.lib.umn.edu/blogosphere/women_ and_children.html.

Herring, S., Scheidt, L., Wright, E. & Bonus, S. (2005). Weblogs as a bridging genre. *Information Technology & People*, *18(2)*, 142-171.

Herring, S. C. (2007). A faceted classification scheme for computer-mediated discourse. *Language@Internet*. Retrieved July 14, 2009 from www.languageatinternet.de/articles/ 2007/2761, *Vol. 4*.

Luzón, M.J. (2006). Research Group Blogs: Sites for Self-Presentation and Collaboration. Paper presented at the V AELFE Conference, Zaragoza, 16 Sept. 2006. Retrieved July 14, 2009 from: http://www.unizar.es/aelfe2006/ALEFE06/5.newtechnologies/87.pdf

Luzón, M.J. (2007). Academic weblogs as tools for e-collaboration among researchers. In N. Kock (Ed.), *Encyclopedia of e-collaboration*. Idea Group Inc., Nueva York. pp. 1-6.

Luzón, M.J. (2008). Social presence in academic blogs. Paper presented at the *Interlae International Conference. Interpersonality in Written Academic Discourse: Perspectives Across Languages and Cultures*. Zaragoza, 2008, December 11-13.

Maingueneau, D. (2008). Modes de généricité et hypergenre. Paper presented at the *International Seminar Genre and ICT: New Competencies and Applications to Autonomous Language Learning*. Castellón, 2008, November.

Mortensen, T. & Walker, J. (2002). Blogging thoughts: Personal publication as an online research tool. In A. Morrison (Ed.), *Researching ICTs in Context* (pp. 247-279). Oslo: InterMedia Report.

Nardi, B. A., Schiano, D. J., Gumbrecht, M. & Swartz, L. (2004). Why we blog. *Communications of the ACM*, 47 (12), 41-46.

Paquet, S. (2002). Personal knowledge publishing and its uses in research. Retrieved July 12, 2009 from http://radio.weblogs.com/0110772/stories/2002/10/03/personalKnowledge PublishingAndItsUsesInResearch.html

Paulus, T. & Dennen, V. (2005). Weblogs as spontaneous mentoring mechanism for academics. *The Association of Internet Researchers 6th Annual Conference*. Chicago. 6-9 October.

Schilt, M. (2006). Is the Future of Legal Scholarship in the Blogosphere?. Retrieved June 30, 2009 from http://www.law.com/jsp/llf/PubArticleLLF.jsp?id=1188464547361

Schmidt, J. (2007). Blogging practices: An analytical framework. *Journal of Computer-Mediated Communication*, *12(4)*, article 13. Retrieved July 12, 2009 from http://jcmc.indiana.edu/vol12/issue4/schmidt.html

Torres-Salinas, D., and Cabezas-Clavijo, A. (2008). Los blogs como nuevo de comunicación científica. In *III Encuentro Ibérico de Docentes e Investigadores en Información y Documentación*, Salamanca (Spain), 5-7 May 2008.

Walker, J. (2006). Blogging from inside the ivory tower. In A. Bruns & J. Jacobs (Eds.), *Uses of Blogs* (pp. 127-138). New York: Peter Lang.

Zivkovic, B. (2006). Publishing hypotheses and data on a blog - is it going to happen on science blogs? Retrieved July 14, 2009 from http://sciencepolitics.blogspot.com/2006/04/publishing-hypotheses-and-data-on-blog.html

APPENDIX I: ANALYSED BLOGS

Aetiology (http://scienceblogs.com/aetiology/). A blog by T.C. Smith, an Assistant professor of Epidemiology.

Beck Institute Blog (BIB) (http://www.beckinstituteblog.org/), the blog of the Beck Institute for Cognitive Therapy and Research.

Cosmic Variance (http://cosmicvariance.com/). A group blog written by a group of physicists and astrophycists.

Culture cat: Rhetoric and Feminism (http://culturecat.net/). A blog by Clancy Ratliff, an assistant professor in the Department of English at the University of Louisiana.

*Developing Intelligence (DI) (*http://scienceblogs.com/developingintelligence/).

Life of a PhD student (http://lifeandtimesofaliteraturephdstudent.blogspot.com/)

Nmrg (http://spooner.beds.ac.uk/nmrg/), the weblog of the new media research group of the Research Institute for Media, Art and Design at the University of Bedfordshire, UK.

Princeton S Network Systems (SNS)* group blog (http://sns.cs.princeton.edu/blog/)

Real Climate (http://www.realclimate.org/). A group blog on climate science by working climate scientists.

Rudd Sound Bites (RSB), the blog of the Rudd Center for Food Policy and Obesity at Yale University (http://www.ruddsoundbites.typepad.com/)

In: Internet Issues: Blogging, the Digital Divide and Digital... ISBN: 978-1-60741-824-5
Editor: Cassie M. Evans, pp. 71-82

Chapter 4

BLOGGING IN HIGHER EDUCATION

[1,2]Patricia D. Wolf and [3]Richard W. Smith

[1]University of Maryland University College, Adelphi, MD, USA
[2]Shippensburg University, Shippensburg, PA, USA
3Harford Community College Bel Air, MD, USA

ABSTRACT

A "blog" is defined by Merriam-Webster's online dictionary as "a Web site that contains an online personal journal with reflections, comments, and often hyperlinks provided by the writer; also : the contents of such a site." According to Encyclopaedia Britannica online, "Some blogs operate mainly as news filters, collecting various online sources and adding short comments and Internet links. Other blogs concentrate on presenting original material. In addition, many blogs provide a forum to allow visitors to leave comments and interact with the publisher." "To blog" is the act of composing material for a blog. Materials are largely written, but pictures, audio, and videos are important elements of many blogs. The "blogosphere" is the online universe of blogs. Common blogging sites that offer users the ability to create blogs for free include LiveJournal, WordPress, and Google-owned Blogspot.com. Other commonly known blogs and related social websites are Tumblr, MySpace, Facebook, and LinkedIn. The microblogging site Twitter, in particular, has become a media sensation and is creating its own role in society and subsequently in education. Blogs are being used in higher education in many ways: as places faculty exchange ideas, by libraries posting news and events, as recruitment tools, as teaching tools, and by faculty to converse with students outside the classroom. However, there is dispute about the impact of blogs in higher education. This chapter reviews the ways in which blogging is being used in higher education and the debate about the impact of such uses.

INTRODUCTION

"A blog--a shorthand term that means 'Web log'--is an online, chronological collection of personal commentary and links" (Educause Learning Initiative, 2005). A "blog" is defined by Merriam-Webster's online dictionary (2009) as "a Web site that contains an online personal journal with reflections, comments, and often hyperlinks provided by the writer; also : the contents of such a site." According to Encyclopaedia Britannica online (2009), "Some blogs operate mainly as news filters, collecting various online sources and adding short comments and Internet links. Linking to these related pieces of information is becoming increasingly common. Other blogs concentrate on presenting original material." "Most weblogs have features that enable readers to comment on individual posts and to comment on the comments, thus creating conversations among participants" (Rosen & Nelson, 2008, p. 216). The majority are maintained by individuals. "Most have only a few readers, but some have readers numbering hundreds of thousands" (p. 216). "To blog" is the act of composing material for a blog. Materials are largely written, but pictures, audio, and videos are important elements of many blogs (Rosen & Nelson, 2008). The "blogosphere" is the online universe of blogs.

Blogs (e.g., LiveJournal, WordPress, and Google-owned Blogspot.com) and their related social networking sites (e.g., Tumblr, MySpace, Facebook, and LinkedIn) are considered the backbone of what is known as Web 2.0. "Web 2.0 is the next phase of Internet usage. The first phase, Web 1.0, focused on presenting information. The next phase, Web 2.0, enables both presentation and participation" (Rosen & Nelson, 2008, p. 212). Web 2.0 enables the Internet to be interactive and makes online collaboration possible (Lemley & Burnham, 2009). "Web 2.0 tools … allow users to transform individual activities … into social-sharing activities by facilitating the creation of Internet-based communities joined for a common purpose, providing these communities with new, on-the-fly ways to publicly share and discuss content … and requiring much less technical knowledge than their predecessor Web 1.0 tools" (Rosen & Nelson, 2008, p. 213). "Social networking sites are ... places on the internet where people converge around shared interests or causes" (Supe, 2008, p. 120). The microblogging site Twitter, in particular, has become a media sensation and is creating its own role in society and subsequently in education.

Web 2.0 does not necessarily signify an advance in web technology. Many Web 2.0 sites use technology that has been around since the early days of the World Wide Web. Instead it signifies a newer and more interactive expression between website and its users. Technology is taking a backseat to interactive communication.

Tumblr, launched in 2007, "lets you effortlessly share anything. Post text, photos, quotes, links, music, and videos, from your browser, phone, desktop, email, or wherever you happen to be" (http://www.tumblr.com/about). This site, along with similar sites such as Soup.io, are called Tumblelogs. It is possible to host a more traditional blog on a Tumblelog, but it is designed more to be a place for the blog authors to quickly display items that personally interest them, often with very little explanation.

MySpace, launched in 2003 and considered among the first of the social networking sites, promotes itself as way to meet people. Facebook, launched in 2004 and generally considered the leading social networking site among college students, is "a social networking site where users interact through a constantly evolving set of networks based on college or university, friendships, interest groups, favorite movies, and other criteria" (Educause Learning Initiative, 2007a). MySpace, now owned by Rupert Murdoch's News Corporation, bills itself primarily as a place to meet people; Facebook provides a way to connect with people you already know (Ostrow, 2009a).

LinkedIn differs from Facebook and MySpace in that it is "more geared to professional networking ... in a context where you are clearly indicating that you are looking for professionals who can assist you, not online friends" (Supe, 2008, p. 119).

Twitter, introduced in 2006, is "an online application that is part blog, part social networking site, part cell phone/IM tool, designed to let users answer the question, 'What are you doing?' Users have 140 characters for each posting (or 'tweet') to say whatever they care to say" ("Educause Learning Initiative, 2007b). "Unlike a formal blog, Twitter doesn't require that users stick to a specific theme" (Kroski, 2008). In 2008, use of Twitter grew by 1,382 percent worldwide (Whittock, 2009). Traffic to Twitter doubled between February and April 2008 (Kroski, 2008). "In March 2009 Nielson Online reported that Twitter has now surpassed Facebook to become the fastest-growing social networking site--ever." Twitter is a "microblogging" site. Microblogging differs from traditional blogs mainly due to its 140-character limit and lack of embedded multimedia. The Laconi.ca project is an interesting open-source variation on the closed platform Twitter.

YouTube is "a user-generated video-sharing community" (Allocca, 2008a, p. 8). Users can post videos of just about anything to the community for others to view, comment on, and rate.

"Wikis are the premier tool for collaboration. A wiki is a Web site that can be edited online by anyone with access privileges and a Web browser. Unlike weblogs in which content is added chronologically but not usually changed, the content and navigation of wikis can be added to, deleted, or edited. … One of the best known wikis is Wikipedia. Because anyone can edit it, the reliability of its information has been suspect, and using Wikipedia has been controversial." (Rosen & Nelson, 2008, p. 217). "Wikis have discussion and commenting features that allow participants to talk about their particular entries, why they made the edits they did, and what their thoughts are … about other participants' edits. … Wikis also have a history roll that allows participants to see every change that has been made and to compare those changes" (Rosen & Nelson, 2008, pp. 218-219).

Social networking sites allow ways for people to connect. "There are three key aspects to social networking that sustain user interest: a sense of community, the development of friendships, ease with which they can interact with other users and ultimately learn from them" (Supe, 2008, p. 121).

Demographics

Twitter is primarily viewed as being used by people over the age of 25 (Martin & MacDonald, 2009) or even 35 (Mansfield, 2009); however, its "buzz" is growing, and there are some data that its use among younger viewers is growing (Ostrow, 2009b).

"Recent studies have shown that 30 percent of Facebook users and 32 percent of MySpace users are older than 45" (Harris, 2008). "A recent Nielson report on twitter reveals that the twitter population skews heavily in the direction of over-25-year-olds, with teens making up less than 16% of the twitterverse" (Martin & MacDonald, 2009). Most users are likely to assume that social network sites are for younger people, but these statistics are showing that the opposite is true. In contrast, "YouTube is where an incredible 13.2 million young people under the age of 18 congregate, and where *another* 13.2 million 18- to 34-year-old users visit every single day" (Allocca, 2008a, p. 8). Sites that are enhanced with video skew to the younger demographic, while the text based sites such as Twitter click with the older population.

Use of social networking sites does vary by ethnicity. "Among different racial and ethnic groups, Hispanic students spent the least amount of time on social-networking sites, with 33.4 percent spending less than one hour per week" (Hoover, 2008). This fact proves that social networking sites, although increasingly important to colleges, must not be the only tool for reaching students. Colleges have to know and understand their audience.

Although it is unclear what the demographics are currently on each of these sites and they seem to change rapidly, it is important to understand what the potential audience and users are on a site before using it.

Use by Faculty

Social networking tools are being used by faculty in a number of ways:

- For recapping class discussions (Rosen & Nelson, 2008). A blog or other social tool is a quick and simple method to summarize these discussions in a simple and linear method. Twitter, with its 140 character limit, makes an interesting tool in that unwanted ideas are likely dispensed with.
- Directing students to class-related online items of interest (Rosen & Nelson, 2008). Social bookmarking sites such as delicious.com or digg.com may be used to link to these related resources. The students may then participate in additional discussions or voting on the significance of the outside link.
- To promote student collaboration and discussion (Rosen & Nelson, 2008; Skiba, 2008). It has been shown that younger students may be used to communicating in a social manner. Providing them online tools to enhance communication in this manner could only serve to increase the overall level of participation in class. A recent study by the University of Minnesota found that "low-income students are just as technologically proficient as their counterparts and credit [social networking sites] for teaching them technology skills, as well as creativity, and providing exposure to diverse views" (Harris, 2008).
- For student portfolios (Rosen & Nelson, 2008). Social tools such as slideshare.com or Google Docs allow students to share their materials with their classmates in an online environment. This allows for collaborative activities.
- To foster a sense of community among students (Kroski, 2008). Younger students may find the traditional classroom is a difficult environment for a community to develop, whereas the social networking sites seem like a community to them.
- To communicate ideas in real time from conferences (Skiba, 2008; Young, 2009)
- To communicate important changes to class information (e.g., class cancellations) (Skiba, 2008). A blog or other social site is an effective method to quickly get important facts out to the students.

- For students to post quick questions about assignments (Skiba, 2008). Once again the 140-character limit of Twitter forces quick questions from the students and also forces equally quick responses from the instructor.
- To follow a public figure on a site such as Twitter and see how often the subject being studied is mentioned (Skiba, 2008). Celebrities and politicians are quickly embracing social networking sites and posting numerous facts about their life and work.
- News feeds can be used by a class to "subscribe to everyone else's weblogs, thus keeping up to date with the conversation going on" (Rosen & Nelson, 2008, p. 219). "News feeds can also be created from search engines, bringing the most current information—text, images, audio, or video—on a topic, a useful research tool for class studies and projects" (Rosen & Nelson, 2008, p. 219).
- To feel connected to or network with far-flung colleagues (Whittock, 2009; Young, 2009) or to share teaching tips or information about teaching and education (Young, 2009).
- As a Web- or news-monitoring resource (Whittock, 2009; Young, 2009). Faculty have found "following" leaders in their field or news organizations (e.g., on Twitter) to be an excellent way of keeping up with developments in their field.

"The term Web 2.0 students identifies a new generation of learners who are comfortable with and enthusiastic about using collaborative technologies to participate in the World Wide Web as creators rather than consumers. These students gravitate toward group activity, seeking interaction within thriving online communities of generative individuals" (Rosen & Nelson, 2008, p. 200). Rosen and Nelson (2008) caution, "of course, participation per se does not equal learning. However, through participation, students engage more in a course, reflect more on course material, and thus are repeatedly exposed to course concepts" (p. 217). The wiki history tool "can help instructors see how students are progressing through a project, which students are contributing which parts to a project, and which areas are not clear to students, thus allowing instructors to clarify those areas and tailor their feedback to the students" (Rosen & Nelson, 2008, p. 219).

Although many faculty are experimenting with social networking sites in their classes, some are still uncertain as to the benefits of such tools, and many express the desire for more training on the possible uses of Web 2.0 tools (Lemley & Burnham, 2009). Although it has been shown that many of the social networking sites have an older demographic, older faculty seem to be reluctant to work within these types of Web tools.

"Since we all have limited time and resources, we should take a hard look at the tools we are using today to manage our professional relationships. Are you able to give them enough time and attention and are the relationships connecting you in with the right people in the right way? ... As with everything, you get as much out of your networking tools as you invest; so allocate your time and resources wisely" (Supe, 2008, p. 120). Spending time on a social tool that does little to help professionally is not necessarily wasted time, but it is not a good allocation of resources if it does not profit the user.

Use by Higher Education Institutions

Many colleges and universities are jumping on the social networking bandwagon. "The University of Massachusetts Dartmouth Center for Marketing Research recently conducted one of the first statistically significant, longitudinal studies on the usage of social media by college admissions offices. The new study compares adoption of social media between 2007 and 2008 by the admissions offices of all the four-year accredited institutions in the United States. The colleges and universities were identified using a directory compiled by the University of Texas" (http://www.umassd.edu/cmr/studiesresearch/mediaandadmissions.cfm). Familiarity with social networking sites increased from 55 to 63%; use of those sites rose from 29 to 61%. Use of blogs went from 33 to 41%.

Ways in which social networking is being used by higher education institutions include the following:

- Helping alumni find jobs using LinkedIn (Lavrusik, 2009)
- Connecting alumni and current students using Facebook groups (Lavrusik, 2009)
- Fundraising using Twitter or Facebook (Lavrusik, 2009)
- Training alumni to use social media (Lavrusik, 2009)
- Engaging alumni through mainstream sites (Twitter, Facebook, LinkedIn) instead of proprietary networks (Lavrusik, 2009)
- Sharing information with alumni from school's website through use of a widget (Lavrusik, 2009)
- Encouraging alumni to generate their own content through wikis and photo-sharing sites (Lavrusik, 2009)
- Promoting official groups on mainstream social media sites on school's alumni page (Lavrusik, 2009)
- Developing mobile applications for use during reunions (shuttle schedules, agendas) (Lavrusik, 2009)
- Using Google Maps to show location of alumni (Lavrusik, 2009)
- Finding friends with similar interests (Allocca, 2008b)
- Posting news stories about your school (Allocca, 2008b)
- Using community bulletin boards (Allocca, 2008b)
- Starting private groups and use them as communities for collaboration (Allocca, 2008b)
- Creating events (Allocca, 2008b)

Many universities and colleges are using social networking sites to solicit new students (Harris, 2008). Social networking sites are free, available to anyone with an Internet connection, and are where current and prospective students are spending their time. "This is where your admissions reps should be. This is where your financial reps should be" (Allocca, 2008b, p. 6). Content is ultimately more important than the format. Many of the most popular social networking tools and applications are in formats that are lacking style, but are popular because of their content.

Because YouTube's audience is the target age for recruiting students, many higher education institutions are considering it as a recruiting tool in the following ways:

- InVideo ad: runs for 15 seconds into a current YouTube video clip. Include the school's contact information at end.
- Branded channel: a custom profile with the school's own URL that links back to its main Web site
- In-line advertising: one or two lines of text to promote the school. Appears at bottom of YouTube video. Text is hyperlink to the school's Website.
- A contest: Prominently featured on the YouTube site (Allocca, 2008a).

Other ways to use video include:

- Become a virtual student
- Tour a dorm or other building on campus
- Film interviews with star athletes
- Film courses
- Create parodies of popular TV shows (e.g., CSI by University of South Carolina Aiken) (Allocca, 2008a, p. 8).

If you are using videos, keep them short: "15 to 30 seconds is perfect for today's attention spans" (Allocca, 2008a, p. 8). Be sure to market your video: embed it on your Website or blog, send out an email blast, post it on the school's Facebook or MySpace site. "The more different and interesting the video is, the more it will be viewed and shared" (Allocca, 2008a, p. 8). The stories you post on social networking sites should be personal experience and should "leave an impression" (Allocca, 2008a, p. 8).

Wilmington University created an application that "combines CraigsList apartment listings with Yahoo maps, so that students looking for housing can see what is currently available and the location's proximity to campus. Students can also use the group to find roommates" (Lorenzetti, 2009, p. 5). Second Life is being used to recruit and retain students, teach classes, and as supplementary learning experiences. It allows for "role-playing, collaboration, real-time interactions between students and faculty, and experimentation" (Skiba, 2009, p. 129). "Online socializing … is a normal, everyday, and familiar way for current and prospective students to congregate and receive information" (Allocca, 2008b, p. 6). College administrators often do not understand the significance of this form of socializing for the students. They may not see it as a normal form of communication. They may also view it as unneeded or unprofessional.

Many colleges and universities have set up accounts on Twitter—with mixed results. Given the demographics on Twitter, the most successful campaigns are ones that engage alumni. "Why? Because they all fall in the right demographic for Twitter! … Broadly speaking, the university personnel to whom we refer fall directly in the target demographic for Twitter use. ... But … we need to do a better job of understanding the audience on Twitter and catering communication effectively. For example, planning an alumni engagement strategy on Twitter is a great direction to move in. Your target group is older, more likely to engage on Twitter, and less likely to be consistently getting information about the school from other places" (Woods, 2009). Colleges need to realize that the demographics of Twitter may not fit their recruiting needs. Twitter may be the hot website of today, but if it does not fit a college's needs, it should not be used. If a college desires this older demographic, then Twitter is a perfect tool, but it may not be for younger students.

Mansfield (2009) contends, "the vast majority of colleges and universities are still struggling with Twitter conversation etiquette and getting past the news pushing component of participating in the 'Twitterverse'" (p. 27). Her 10 tips for using Twitter effectively:

1. Put authenticity before marketing. Have personality. Build community.
2. Don't use Twitter for RSS or publish "News" unless you call your Twitter profile "News." News is not authentic and lacks personality.
3. Have many Twitter accounts. Have separate accounts for news, athletics, admissions, alumni.
4. Be nice. Be thankful. Reply and retweet.
5. Follow everyone who follows you. How can you build community on Twitter if you won't even participate with your followers?
6. Use "favorites" to organize the chaos and feature your most important tweets.
7. Don't tweet about your coffee, the weather, or how tired you are. Provide value to your followers, not chitchat.
8. Don't only tweet your own content. Tweet articles or blog posts by your favorite newspapers, bloggers, or nonprofit organizations.
9. Send messages, but not via auto-responders. Respond to specific questions or messages you've received. Don't send spam.
10. Limit your tweets to five per day and definitely post no more than six.

Other ways universities and colleges are using Twitter include the following:

- Libraries use Twitter to post about new acquisitions, events at the library, or to post and answer reference questions (Kroski, 2008; "Messaging in a 2.0 World," 2007).
- Universities post news about the institution but also thoughts about trends in social media (Young, 2009)
- The University of Illinois has an impressive 36 Twitter feeds—everything from Admissions and Alumni offices to personal feeds by athletic coaches (http://illinois.edu/about/overview/cool/twitterfeeds.html). The University of Buffalo has more than 20 Twitter feeds (http://crowdstatus.com/UB% 20on% 20 Twittercrowd.aspx).

CONCERNS ABOUT SOCIAL NETWORKING SITES

Some writers are concerned about the potential omnipresence of social networking sites: "With more institutions moving toward commercial sites for student e-mail, calendaring, and document applications, it should not be too long before higher education collaborates with Facebook to link applications for course enrollment, grade checks, and other online student services" (Mitrano, 2008). In fact, Blackboard already offers the Learn application (http://www.facebook.com/apps/application.php?id=27522527824) that provides course updates from within Blackboard Academic System. According to Mitrano (2008), "a corporate, commercialized Internet has more money, flexibility, and motivation to innovate than do most business aspects of higher education and is the driving force behind the outsourcing of campus IT services and products." Mitrano (2008) cautions, "IT professionals—vice presidents and chief information officers especially—have a responsibility to raise critical questions and perhaps even to teach or coach their administrations about the long-term, and possibly unintended, deleterious consequences of decisions that seem so obvious from a business and financial perspective today."

Other authors caution against the idea that students savvy in social networking have exceptional computer skills: "In our own experience, we have observed that many of today's students who have heard of and even used MySpace and Facebook had not previously heard the term blog; some knew of Wikipedia, but none knew what a wiki was; and none had heard of RSS before. Moreover, their skills with video games, TV, and cell phones do not readily transfer to other applications. Thus, in introducing Web 2.0 tools into the educational process, it is important to keep in mind that our students are not a homogenous group of digitally enhanced students" (Rosen & Nelson, 2008, p. 221). A student may be able to text with ease, but struggles with basics of computer applications.

However, there are also dangers in not being familiar with these new media. Timothy L. Wood, an assistant professor of history at Southwest Baptist University in Bolivar, Missouri, had his online identity usurped. He was mistakenly credited with writing an essay on an extremist Web site comparing President Barak Obama to Adolf Hitler. The essay was subsequently amended to include Professor Wood's affiliation, office phone number, and email address. Professor Wood was outraged: "Who had the right to speak for me like that? How dare they hide behind my name! What if my colleagues -- or my friends and family -- read this and believed it?" (Wood, 2009).

"But the most pressing question seemed to be what kind of damage control would be necessary in order to prevent this from irreparably damaging my career. And that, in turn, led me to begin reflecting on how scholars will need to safeguard their professional reputations in the 21st century. Although I would never wish this kind of ordeal on anybody, the realist inside me fears that I will not be the last professor to fall victim to digital dishonesty. As academics, we must be aware that our professional reputations are transmitted through the technology of a bygone era, and even then are typically shrouded in secrecy or obscurity. Mentors, colleagues, and administrators exchange sealed and confidential references printed out on university letterhead. Editors, referees, and reviewers validate our scholarly work by allowing us access to or giving us coverage in their publications, but the results of that process all too often lie buried in library stacks and academic databases. In the meantime, the malicious or misinformed denizens of the Web have had time to hit the 'forward' button about a million times." Professor Wood concludes that all academics should at least Google their name and be aware of what might be posted about them on the Internet (Wood, 2009).

There have also been questions raised about whether administrators or professors should be or are even welcomed in the perceived student-only spaces of social networking. "Both students and faculty members have mixed reactions. Though faculty-student interaction on social networking sites encourages relational exchange, issues of self-disclosure and identity management are also of concern. How much is too much? How much information should a faculty member or student share with each other? Faculty fear of losing credibility is also germane. There is the potential violation of student expectations, and university and administrative expectations of proper behaviors" (Harris, 2008). "Building interpersonal relationships among students, academicians and administrators has the potential to alter perceived power relationships by making faculty and personnel seem more accessible" (Harris, 2008). There needs to be a balance between being accessible for the students, but yet still keep a proper academic presence.

Students also have been affected by what they post on social media sites. The University of Texas removed a student from the football team after he posted a racial slur on his Facebook page (Debolt, 2008). Calvin College in Grand Rapids, MI, expelled a student for one year after he allegedly posted a lewd comment about his ex-girlfriend on Facebook (Nealy, 2009), and John Brown University administrators expelled a student when photos of him dressed in drag appeared on Facebook, saying the student had violated the private, Christian college's code of campus conduct (Nealy, 2009). Such incidents do still remain rare, though, considering the numbers of students who use social media sites.

"Documenting ... behavior online is a recent phenomenon that still vexes administrators. ... When social-networking sites first became popular, many colleges and universities watched them closely. Since then, a consensus has emerged that institutions should not actively monitor the sites or react hastily to information they discover. Such practices run counter to sound educational philosophies, officials say, and make colleges vulnerable to lawsuits. ... Administrators have a responsibility to investigate if they learn of online evidence that a student may have violated their college's conduct code, or of postings that may be offenses themselves. ... [but] if colleges actively monitored Web sites like Facebook and MySpace, they would set an expectation--and a legal duty--to do so effectively and consistently. ... And administrators realize they cannot ignore reports of misconduct online. Even if they do not actively monitor social-networking sites, a disturbing post brought to their attention puts them on notice to respond. If they don't, they may be found negligent in court. ... But some student groups still warrant extra scrutiny. ... Resident advisers, for example, are employees of a college and may be held to stricter standards of behavior. The same applies to athletes and to students with discretionary appointments, such as to honor councils. Their coaches or supervisors may reserve the right to examine their online profiles and ask them to remove offensive material, or revoke their special privileges. ... Whatever standards colleges adopt, they should be made clear to students" (Lipka, 2008). "In the future... institutions will need to develop social networking policies that outline for students what is appropriate and what is not" (Nealy, 2009).

CONCLUSION

Blogs vary in the ways in which they present material, how they can be edited, and how they create community. Blogs of all sorts are being used in higher education in many ways: to recruit new students, to foster community, by faculty within classes, and to connect the institution to donors and alumni. However, the use of blogs is still in its infancy, and many institutions and faculty are struggling with the best way to use social media. Social media blog sites such as Twitter are even newer than "traditional" blogs and their use may add complexity to already conflicted institutions. Some institutions and faculty have even shunned the use of social media, but many authors believe that it is important for everyone involved in higher education at least be familiar with social media and the ways in which it is being used, even if those ways change constantly! Although many are successfully using social media in higher education, there are issues to take into consideration before jumping into this new arena. Social media can be a powerful tool in higher education, as long as it is used appropriately.

REFERENCES

Allocca, J. (2008a). Considering YouTube? *Recruitment & Retention in Higher Education, 22(7)*, 8-9. Retrieved August 7, 2009, from Academic Search Premier database.

Allocca, J. (2008b). Student marketing 2.0: tapping into social networks. *Recruitment & Retention in Higher Education, 22(3)*, 6. Retrieved August 7, 2009, from Academic Search Premier database.

Blog (Internet). (2009). *Encyclopaedia Britannica*. Retreived August, 7, from http://www.britannica.com/EBchecked/topic/869092/blog.

Blog. (2009.) *Merriam-Webster's online dictionary*. Retrieved August 7, from http://www.merriam-webster.com/dictionary/blog.

Debolt, D. (2008). U. of Texas punts lineman after anti-Obama post. *Chronicle of Higher Education, 55(13)*, A13. Retrieved August 7, 2009, from Academic Search Premier database.

Educause Learning Initiative. (2005). 7 Things you should know about blogs. Retrieved August 7, 2009, from http://www.educause.edu/ELI/7Things YouShouldKnow AboutBlogs/156809.

Educause Learning Initiative. (2007a). 7 Things you should know about Facebook II. Retreived August 7, 2009, from http://www.educause.edu/ ELI/7ThingsYouShouldKnow AboutFaceb/156828.

Educause Learning Initiative. (2007b). 7 Things you should know about Twitter. Retreived August 7, 2009, from http://www.educause.edu/ELI/7ThingsYouShouldKnowAbout Twitt/161801.

Harris, K. (2008). Using social networking sites as student engagement tools. *Diverse: Issues in Higher Education, 25(18)*, 40. Retrieved August 7, 2009, from Academic Search Premier database.

Hoover, E. (2008). Colleges face tough sell to freshman, survey finds. *Chronicle of Higher Education, 54(21)*, A1-A22. Retrieved August 7, 2009, from Academic Search Premier database.

Kroski, E. (2008, July). All a Twitter. *School Library Journal,* pp. 31-35. Retrieved August 7, 2009, from Academic Search Premier database.

Lavrusik, V. (2009, July 23). 10 Ways universities are engaging alumni using social media. Retreived July 27, 2009, from http://mashable.com/2009/07/23/alumni-social-media.

Lemley, T. & J. F. Burnham (2009). Web 2.0 tools in medical and nursing school curricula. *Journal of the Medical Library Association, 97(1)*, 49-51.

Lipka, S. (2008). The digital limits of 'in loco parentis.' *Chronicle of Higher Education, 54(16)*, A1-A21. Retrieved August 7, 2009, from Academic Search Premier database.

Lorenzetti, J. P. (2009). Private social networks as part of the admissions cycle. *Recruitment and Retention in Higher Education, 23(3)*, 5-6. Retrieved August 7, 2009, from Academic Search Premier database.

Mansfield, H. (2009). 10 Twitter tips for higher education. *University Business, 12(5)*, 27-28. Retrieved August 7, 2009, from Academic Search Premier database.

Martin, D. & S. MacDonald. (2009, July 30). Tweens don't tweet; Twitter's growth not fueled by youth. *Nielsen News*. Retrieved August 10, 2009, from http://blog.nielsen.com/nielsenwire/online_mobile/teens-dont-tweet-twitters-growth-not-fueled-by-youth/.

Messaging in a 2.0 World: Twitter & SMS. (2007, September-October). *Library Technology Reports*, pp. 62-66. Retrieved August 7, 2009, from Academic Search Premier database.

Mitrano, T. (2008). Facebook 2.0. *Educause Review, 43(2)*, 72-73.

Nealy, M. J. (2009). The new rules of engagement. *Diverse: Issues in Higher Education*, 26(3), 13-31. Retrieved August 7, 2009, from Academic Search Premier database.

Ostrow, A. (2009a, April 13). MySpace welcomes me back; what is MySpace? *Mashable: The social media guide*. Retrieved August 7, 2009, from http://mashable.com/2009/04/13/myspace-marketing-email/.

Ostrow, A. (2009b, September 2). STATS: Young people are flocking to Twitter. Retrieved September 2, 2009, from http://mashable.com/2009/09/02/twitter-demographics/.

Rosen, D. & C. Nelson. (2008). Web 2.0: a new generation of learners and education. *Computers in the Schools, 25(3-4)*, 211-224. Retrieved August 7, 2009, from Academic Search Premier database.

Skiba, D. J. (2008). Nursing education 2.0: Twitter & tweets. *Nursing Education Perspectives, 29(2)*, 110-113. Retrieved August 7, 2009, from Academic Search Premier database.

Skiba, D. J. (2009). Nursing education 2.0: A second look at Second Life. *Nursing Education Perspectives, 30*(2), 129-131. Retrieved August 7, 2009, from Academic Search Premier database.

Supe, A. (2008). Networking in medical education: creating and connecting. *Indian Journal of Medical Sciences, 62(3)*, 118-123.

Whittock, J. (2009, April 30). Twitterati in the academy. *Times Higher Education, 20*. Retrieved August 7, 2009, from Academic Search Premier database.

Wood, T. L. (2009, August 24). The accidental celebrity. Retrieved August 24, from http://www.insidehighered.com/views/2009/08/24/wood.

Woods, A. (2009). If teens don't tweet… then why are universities on twitter? Retrieved August 7, 2009, from http://www.othersidenotes.com/?p=240.

Young, J. R. (2009). 10 High fliers on Twitter. *Chronicle of Higher Education, 55*(31), 10-11. Retrieved August 7, from Academic Search Premier database.

In: Internet Issues: Blogging, the Digital Divide and Digital... ISBN: 978-1-60741-824-5
Editor: Cassie M. Evans, pp. 83-94

Chapter 5

LIBRARIANS AND INFORMATION SCIENTISTS BLOGOSPHERE: AN OVERVIEW

Noa Aharony[*]
Department of Information Science, Bar-Ilan University, Ramat Gan, Israel

ABSTRACT

Blogging, though a relatively new phenomenon, has already gained greatly in popularity, and blogs have become an essential information channel in the Web 2.0 information world. A survey conducted by the Pew/Internet American Life Project (2006) found that 8% of the 120 million internet users in the USA keep a blog, and 39% of users reported reading blogs. These data reveal an increase from a previous Pew/Internet American Life Project (2005) survey, in which only 7% of users reported writing a blog and 27% said they read blogs.

With the growing importance of blogs in our information world, the current research would like to present an overview of library and information science (LIS) blogoshere, focusing on different aspects: LIS professional literature; nature, development and prevailing tendencies in LIS blogosphere; the phenomena of tagging and folksonomies in the LIS blogosphere; and an attempt to understand comments written by readers of LIS blogs. The overview findings are important to those who are interested in blogs. It may cause them to better understand and explore the phenomenon of blogosphere and to better exploit the information assigned in the blogs.

INTRODUCTION: BLOGS

Blogging, though a relatively new phenomenon, is already very popular, and blogs have become a central feature and an essential information channel in the Web 2.0 information world. There are several definitions for the term blog. Merriam Webster Online (2004) sees a blog as "a Web site that contains an online personal journal with reflections, comments, and

* Corresponding author: Tel: +972 – 3-6352173, Fax: +9723 – 5354325, Email: aharonn1@mail.biu.ac.il

often hyperlinks provided by the writer." The site Marketingterms.com (2007) defines blogs by their content as a "mixture of what is happening in a person's life and what is happening on the Web, a kind of hybrid diary/guide site, although there are as many unique types of blogs as there are people." The Webopedia (2004) definition focuses on the personal aspects of a blog: "A blog is a web page that serves as a publicly accessible personal journal for an individual. Typically updated daily, blogs often reflect the personality of the author." Finally, some researchers define a blog as an online journal (Doctorow, Dornfest, Johnson and Powers, 2002).

According to Alcock (2003), blogs came into being in 1998 and were first known as weblogs. The name was shortened to blog in 1999, and the website creator was named a blogger. In the beginning, there were only a handful of blogs because their publication required users to have an extensive knowledge of HTML code. However, with the development of blogging tools that enable users to create their online diaries with ease—without knowing HTML—blog publication grew rapidly. The new blogging tools enable *permalinks* (permanent links to the posted messages) and simple archiving, and they allow readers of postings to comment on messages (Bar-Ilan, 2007). Benkler (2006) suggests that the blog allows a "weighted conversation" between the primary author and his comment contributors.

The literature presents several categorizations of blogs. Mernit (2003) bases her categorization on the owner of the blog, such as a professional journalist or a nontraditional journalist. She also categorizes according to focus, such as bloggers focused on a specific theme, movement, event, topic, or interest; or according to the intended audience, such the education community, the self-expression/journaling crowd, the business/marketing/ promotion community, and experimenters and innovators. Eszter (2003) categorizes the blogs according to different types of styles and contents: (a) a personal journal that describes the author's daily activities, (b) a list of links with commentary by the blogger, (c) interactive commentary on a specific topic, (d) editorial pages (one-way commentary), and (e) hodge-podge blogs. Bar-Ilan (2005) proposes a different classification, and classifies blogs by authorship (single, group, or communal), content (personal or topic-oriented), and format (lists of sites, essays, and/or announcements)—noting, however, that the latter two are not always easily discernable.

The number of blogs is constantly growing and with it, blog research. Researchers have studied blogging in general (Blood, 2004; Herring et al., 2006), the use of blogs by different social sectors (Huffaker and Calvert, 2005; Schaap, 2004), and the use of specific blogs, such as "warblogging" (Thompson, 2003), political blogging (Adamic and Glance, 2005), informational blogging (Bar-Ilan, 2005, 2007) and educational blogging ([Aharony and Bronstein, 2008; Hall and Davison, 2007).

Several studies have focused on a very interesting question: Why do people write blogs? Graham's (2002) list of reasons include the desire to publish, to minimize spam, and to learn about web publishing, as well as the need for a sense of community. Lasica (2005) believes that blogging is a new form of journalism. Turgeon (2004) enumerates different reasons she blogs, including personal, to express feelings and personal understanding; and social, to meet new people and share information with them. She also notes that blogging provides access to nonfiltered sources of information that would otherwise not be accessible via the average media sources. Dvorak (2002) provides the following reasons for blogging: ego gratification, antidepersonalization, elimination of frustrations, a need to share, and a need to publish.

Nardi, Schiano, Gumbrecht, and Swartz (2004) interviewed 23 bloggers on their reasons for blogging and found that they were driven by a desire to document their lives, to share their knowledge and opinions, to express emotions, and to form and maintain community forums.

Exploring the attitudes towards blogs within the Library and Information Science (LIS) profession reveals ambiguous attitudes, despite the recent and general widespread adoption of blogs. Hall and Davison (2007, p.165) write about a range from "plaudits for a serious application of social software to dismissal of frivolous endeavor." LIS professionals understand that blogs provide news to the LIS community and that blogging can be a component of knowledge management because blogs can serve as platforms for knowledge creation. LIS professionals also acknowledge the fact that blogs can serve as a source of competitive intelligence and that they can play a role in monitoring products and services (Haberman, 2005). In addition to this, libraries can use blogs as a form of publication (Maness, 2006), or as tools for marketing their resources and events (Stephens, 2007). Moreover, Bar-Ilan (2007) proposed that library blogs are ideal for disseminating, commenting on, and expressing opinions. However, some LIS professionals, aware of the varying quality of content in blogs, are wary of them.

Tags and Folksonomy

In the last few years, the concept of Internet users creating and aggregating their own metadata has been gaining ground. Tagging services allow participants to freely associate natural language keywords—the tags—to a specific information item (i.e., videos, presentations, music, photographs, a URL, or a news item). The use of tagging is expanding rapidly and it is widely prevalent in social bookmarking and photo sharing sites such as Del.icio.us, Flickr, and Furl (Spieteri, 2007). However, text-based blog posts are also targets for tagging, as bloggers assign tags to their posts (Chopin, 2008). Tagging has been the focus of several recent studies (Aharony and Bronstein, 2008; Dye, 2006; Mathes, 2004; Noruzi, 2006, 2007; Peterson, 2006; Sinclair and Cardew-Hall, 2008; Speller, 2007; Spieteri, 2006, 2007).

Folksonomy has emerged as an alternative to traditional classification. A folksonomy is created as users of the website add tags to describe items on the site. The term folksonomy (from "folk" and "taxonomy") was coined in 2005 by information architect Thomas Vander Wal for the growing phenomenon of user-generated metadata attached to digital pieces of information for retrieval purposes (Dye, 2006). The concept of collaboration is usually associated with folksonomies (Bateman et al., 2006; Fichter, 2006; Golder and Huberman, 2006; Mathes, 2004; Quintarelli, 2005) But, Spieteri (2007) claims that it would be more accurate to say that folksonomies are created in environments where people may not actively collaborate in creating tags, but rather may access and use tags assigned by others. Thus, folksonomies enable the use of shared tags. The visual representation of a folksonomy is called a tag cloud. The size of the words in the tag cloud reflects the number of items that carry that tag. According to Sinclair and Cardew-Hall (2008), a tag cloud is a user interface element commonly associated with folksonomy datasets.

LIBRARIANS AND INFORMATION SCIENTISTS BLOGOSPHERE: RESEARCHES OVERVIEW

Several researches were conducted focusing on LIS blogosphere. This chapter will present several articles dealing with various aspects of LIS blogosphere in order to provide the reader a broader perspective of this professional blogosphere. The first research that will be addressed here (Aharony, 2009a) concentrates on librarians' blogs and describes and analyzes their development, nature and changes. The purpose of the paper was to compare and examine two points in time of the LIS bloogsphere: first period of publication of the blog (first two months for each blog) and a second period (July-August 2008). The primary research question was to explore whether there was a difference in the nature of the blogs in two points in time: the inception of blog and time of study.

In the first phase of the investigation, LIS blogs were located by browsing the top 25 Librarian Bloggers on OEDB (Online Education Database, http://oedb.org/library/features/top-25-librarian-bloggers-by-the-numbers). In addition, LIS blogs were located by searching the term "library and information science" in Technorati, one of the main search engines dedicated to blog search. After reviewing 100 library and information science blogs, 30 were selected based on the following criteria:

1. The central theme of the blog was library and information science. The selected blog had to be topic oriented, dealing mainly with professional issues.
2. The blog had to be active during the two months during which the research was conducted (July-August 2008).

The research was conducted during the summer of 2008 and encompassed 30 LIS blogs. The blog analysis was conducted in two phases: (1) statistical descriptive analysis and (2) content analysis. For the first phase (statistical descriptive analysis) the researcher examined all the posts which appeared in the first period of publication of the blog and in the second (time of research) publication period. For the content analysis the researcher chose four posts for each blog: two from the first period of the blog, and two from the second. Each post had to be at least 100 words long, and 120 posts were included in this content analysis. The four posts from each blog were thoughtfully analyzed.

Findings indicate that in the first period bloggers, who dealt with a new and interesting technological platform, were more enthusiastic. This accounts for the higher number of average daily posts, number of posting days and the fact that the total number of postings in 60 days was also higher. In other words, bloggers in the first period devoted more time and energy to their blogs. This tendency is reinforced in Farkas's blog (2008) who explained that in the beginning she posted at least once a day, and eventually her postings slowed down. This finding may also be associated with the new and spreading phenomenon of microblogging, as more and more people use Twitter and write short messages and not long posts.

But, it should be remembered that although bloggers tended to write less in the second period, there is a change in their pattern of writing. Bloggers write more essay-type posts with hypertext links and less short postings, so that those bloggers who continue writing and adding posts to their blogs prefer the detailed essay-type posting, with links. Moreover, it should be noted that links usually represent different sources of information for bloggers. It seems that along the years, the bloggers were exposed to some of the problematic aspects of Web 2.0 (problems of reliability and authority which are related to user-generated content), which explains the larger number of links embedded in the texts in the second period. Possibly bloggers have understood that in order be reliable, they must substantiate their writing and provide their readers with more examples and illustrations (Salant, 2006). Furthermore, it seems that bloggers have realized that while embedding links to various blogs or other sites, they attract the attention of those bloggers or readers while exposing their own writing to a larger audience (Farkas, 2008; Lee King, 2008). In addition, bloggers assign more tags to their postings than they did in the first period. The bloggers are aware of the tagging phenomenon, including its benefits and weaknesses, and of its contribution to the blogosphere.

Finally, when referring to the comments sent to the blogs, it seems that there was greater reader participation during the first period, when the blogs were first published. Perhaps, the readers too, as the bloggers themselves, were initially more enthusiastic about the blogs and about the possibility of expressing themselves so easily on the web, and as a result in the first period they were more active and wrote more comments.

To sum up, findings demonstrate that despite some major differences between these two periods which may represent new directions in librarians' blogosphere, there are other issues which have not been changed a lot.

Findings show that there is a current tendency to write essay-type posts with hypertext links; there are fewer posts and of postings days, as well as fewer readers' comments, but a larger number of links and tags which are assigned to the posts. These differences may represent a new direction, a change in librarians' blogosphere. But, a thorough examination of blogs' content reveals that three out of the five main content categories are equivalent in both periods and it seems that the same issues still occupy librarians' minds and thoughts. The most obvious finding is that librarians in both periods are mostly concerned with technology and its innovations.

It seems that there are changes and differences which can be characterized as focusing on the format or on the structure as well as on the bloggers' patterns of writing. However, examining the substance of the blogs – its content – reveals no major difference, upheaval or change along the two periods. Librarians' blogs in both points in time convey the bloggers' professional interests and concerns and serve as places for dissemination of professional knowledge and experience with the librarians' community.

The second aspect of LIS blogosphere which will be discussed here will try to understand the tagging patterns of librarian and information scientist bloggers and the nature of the folksonomies they form (Aharony, 2009b). The research portrays a thorough content analysis of tags assigned in 30 LIS blogs and it seeks to explore whether those LIS professionals will use specific, meaningful keywords while tagging their posts, as they usually do in their work, or whether they will use random, causal keywords because blogs are such an informal platform. The two primary research questions were:

1. What is the content of LIS topic-oriented blogs?
2. Do the tags and folksonomies used in these blogs form a consistent scheme? In other words, are the tags organized in a coherent, accurate way (as library catalogues are), or are they chaotic and devoid of internal logic?

The first phase of the investigation was to choose LIS blogs to be included in the sample. The blogs were located by browsing the top 25 Librarian Bloggers, as published by the Online Education Database (2007). In addition, LIS blogs were located by searching “library and information science” in Technorati, one of the main search engines dedicated to blog searches. After reviewing 100 library and information science blogs, 30 were selected, based on the following criteria:

The blog had to be written by a librarian or by an information scientist. The blog had to be active during the entire examination period (May–July, 2008).

The second phase of the investigation was a comprehensive categorization of LIS tags that appear as a tag cloud in the search engine Technorati. Technorati provided 734 tags in total to the 30 blogs. The present categorization includes 455 unique tags, as multiple appearances of the same tag were counted only once.

Referring to the tags categorization and to the question whether the tags and folksonomies form a consistent scheme, it turns out that there are five content categories: general tags, technology-related tags, library-related term, information-related terms, and social web-related tags. The largest category (general tags) in this folksonomy does not reflect bloggers' professional interests but rather the personal dimension of blogging and the folksonomies' power to reflect bloggers' lives. The second largest category (technology-related tags) indicates the significance of technology-related subjects for librarians and information scientists. The third category (library-related terms) reflects the librarians' and information scientists' interest in various aspects of their fields. The following category (information-related terms) shows librarians' and information scientists' interest in aspects of information related to their work life. The smallest category is social web-related tags, which could mean that social aspects of the web are not yet the central theme for librarians and information scientists.

This categorization suggests some significant conclusions. First, four major categories of profession-related content were revealed in the analysis, reflecting librarian and information scientist bloggers' major fields of interest. The second conclusion is that although bloggers seem to randomly assign nonrelated tags to the postings, our content analysis shows that these tags can be categorized.

Summing up, the present research presents an extensive analysis of the tagging on LIS blogs and indicates that the professional LIS blogs reflect the most up-to-date interests of LIS professionals as well as bloggers' personal issues. It is noteworthy; however, that the largest category was the one with general tags (33.62%). This general-tags category reflected bloggers' personal interests and experience and not their professional life. In other words, perhaps the informal nature and platform of blogs enables the professional bloggers to express and share their views about personal issues, even though the platform is a professional one.

The next facet of librarians and information scientists blogosphere will concentrate on examining comments written by readers of LIS blogs (Aharony, 2009c). In most blogs readers can comment on the postings, thus engage in active discussion, and become part of the blogosphere without actually setting up a blog. The following research describes and analyzes the comments assigned to the posts, focusing on the language of the comments, the type of information presented in the comments, and the content of the comments. The primary research questions were:

1. What type of language do comment writers use?
2. What type of information is found in the comments?
3. What is the content of the comments?

The research presents an analysis of comments which appeared in 30 LIS blogs in August-October 2008. The researcher examined the comments assigned to the posts and the analysis was conducted in two phases: (1) statistical descriptive analysis and (2) content analysis. For both phases, the researcher examined all the comments which were assigned to the posts during the examined period.

In the first phase of the investigation, LIS blogs were located by browsing the top 25 Librarian Bloggers on OEDB (Online Education Database, http://oedb.org/library/features/top-25-librarian-bloggers-by-the-numbers). In addition, LIS blogs were located by searching the term "library and information science" in Technorati, one of the main search engines dedicated to blog search.

The second phase of the investigation was a statistical descriptive analysis. The third phase was a content analysis which was divided into three themes and created a categorization of writer's language, different types of information presented in the comments and comments' content.

Exploring the first part of the content analysis reveals that most of the comments were written in a personal language, a phenomenon which is a natural and expected one, as the readers add their own personal experiences or thoughts to the posts. The second part of the content analysis which focuses on the different types of information conveyed in the comments shows that most of the comments included substantive information. In other words, the comments' writers did not express only courtesy comments but they took advantage of the platform and expressed personal, impressive, advisory, as well as reflective comments. These findings are similar to Farkas's (2007) findings, that most blogging librarians consider their blogs as personal, although many wrote posts about professional topics. In the current study, the comment writers used personal style and wrote about personal issues as well as about professional ones. It seems that the comment writers chose the blog as a place where they can share their thoughts, impressions, and tips with other readers. Writing the comments, provided them with an opportunity to use higher levels of cognition, to discuss and reflect issues which most concern them. These findings direct us to the third part of the content analysis which concentrates on the actual content of the comments. Although the largest category deals with general issues, the other categories which are quite large (except the last two) focus on library, technology, Web 2.0, general librarianship, and open access and conferences. All these issues are professional and reflect librarians' fields of interest. Obviously, the LIS blog readers who participate and add comments to the posts took advantage of this platform and discuss topics which are relevant to librarians' work and profession. The writing of the comments enables them to rethink, reflect, evaluate, and reconsider professional issues and to share it with their colleagues. These findings echo Aharony (2009b) and Aharony and Bronstein (2008) who explored LIS blogs and education technology blogs and found that bloggers in both areas write about professional issues. Thus, the current findings add a new dimension to the LIS blogosphere. It is not only the LIS bloggers who deal with professional issues in their blogs, but it is also the readers who add comments, write, and discuss professional topics. It seems that both the bloggers and their readers exploit this medium for dissemination of professional information as well as for sharing their experiences and thoughts.

The last aspect of LIS blogosphere which will be discussed in this chapter will delve into the issue of whether the phenomenon of blogs is significantly expressed in the professional LIS literature over the three past years (2006-2008) in Library and Information Science Abstracts (LISA) database. Furthermore, it will display which kind of journals deal with blogs (Aharony, 2009d). The LISA database was chosen based on the premise that contemporary professional literature, as presented in bibliographic databases, exposes the fullest, most current, most objective, most intelligible indicators of the examined subject (White, 2001).

The whole paper which focuses on the subject of Web 2.0 and its main applications: blog, wiki, social network and tags, presents a statistical descriptive analysis and a thorough content analysis of descriptors and journal titles extracted from Library and Information Science Abstracts (LISA) database. The research was conducted in December 2008 and contained two phases. The first focused on the descriptors. This analysis was conducted in two stages: (1) statistical descriptive analysis and (2) content analysis. For the statistical and descriptive analysis the researcher examined all the descriptors assigned to Web 2.0 issues and its main applications. The second phase of the research addressed the journal titles, and included a separate content analysis. The following analysis will concentrate only on blog application.

The primary research questions concerning our issue are presented below:

- Whether the phenomenon of blogs is significantly expressed in the professional LIS literature;
- Which kind of journals deal with blogs.

The research findings indicate that the total number of articles focusing on Web 2.0 and its main applications was 472. The highest number of articles (135) was associated with blogs (28.60%). Furthermore, the lowest number of peer- reviewed articles is associated with the blog descriptor (5.18%).

Blogs seem to be the main interest of the writers. But a thorough examination of the peer-reviewed articles according to their different applications reveals that only 5.18% of articles on blogs are peer-reviewed. These findings suggest that there are many non professional or non scholarly articles focusing on the issue of blogs. The blog is a phenomenon that attracts the attention of a variety of people, and thus different popular articles deal with it. It is noteworthy referring to the largest number of "less academic" articles which deal with blogs. A possible explanation for this phenomenon may be that the LIS community is built of practitioners as well as of scholars and researchers. These articles reveal a practical, how to do aspects which are useful and necessary for librarians and information scientists. The practical articles may direct the practitioners how to use, build, promote and assimilate blogs into their libraries.

In order to delve into the second research question, which focuses on the kinds of journals dealing with the blogs phenomenon, a content analysis of journal titles was conducted. This analysis works on the assumption that journal titles reflect the main issues expressed in the journal; the current study sought to examine which kind of journal is familiar with the blogs phenomenon and publishes articles dealing with it. According to the content analysis that was carried out, a further interesting finding shows that the highest percentage of journal titles that mention blogs (31.11%) is associated with library topics. Perhaps the editors of these journals understand the potential that lies in blogs as it may reflect new tendencies in the libraries. Furthermore, it seems that they grasp that exposing this new tool to the LIS community may cause librarians to assimilate it in their workplace, and as a result library image can be improved.

CONCLUSION

Summing up, this chapter has delved into LIS blogosphere and revealed some interesting facts about this particular professional blogosphere. This paper may enable librarians and information scientists to better understand, explore and participate in the LIS blogosphere. Furthermore, the findings described in this chapter, shed light on the importance of topic oriented blogs as they deal with both professional as well as with personal information, and provides the professional community an opportunity to exploit the information assigned in the blogs.

REFERENCE

Adamic, L. & Glance, N. (2005*). The political blogosphere and the 2004 US election: divided they blog*. Paper presented at the WWW2005 Blog Workshop. Retrieved March24, 2008 from: www. blogpulse.com/papers/2005/AdamicGlanceBlogWWW.pdf.

Aharony, N. (2009a). An exploratory analysis of librarians' blogs: Their development, nature and changes. *Aslib Proceedings*, *61(6* 587-604).

Aharony, N. (2009b). Librarians and information scientists blogosphere: An exploratory analysis. *LISR* 31, 174-181.

Aharony, N. (2009c). *LIS Blogs Comments: An exploratory analysis. Libri. 60*, **(1)**, 65-77.

Aharony, N. (2009d). *Web 2.0 in the professional LIS literature: An exploratory analysis. Journal of Librarianship and Information Science*, In press.

Aharony, N. & Bronstein, J. (2008). The educational technology blogosphere: An Exploratory Analysis. Paper presented at ASMI conference, Haifa, Israel.

Alcock, M. (2003). *Blogs – What are they and how do we use them? Quill, 103* (8). Retrieved February, 24, 2008 from: alia.org.au/groups/quill/issues/2003.8/blogs.html

Bar-Ilan, J. (2005). Information hub blogs. *Journal of Information Science 31(4)*, 297-307.

Bar-Ilan, J. (2007). The use of Weblogs (blogs) by librarians and libraries to disseminate information. *Information Research, 12(4).* Retrieved March, 2, 2008 from: http://InformationR.net/ir/12-4/paper323.html.

Bateman, S., Brooks, C. & McCalla, G. (2006). *Collaborative tagging Approacjes for ontological metadatain adaptive e-learning systems.* Retrieved May, *24*, 2008 from: www.win.tue.nl/SW-ElL/2006/camera-Readyo2/bateman_brooks_McCalla_SWEL 2006. final.pdf.

Benkler, Y. (2006). *The wealth of networks: How social production transforms markets and freedom.* Yale University Press: USA.

Blood, R. (2004). How blogging software reshapes the online community. *Communications of the ACM, 47(12)*, 53-55.

Chopin, K. (2008). Finding communities: Alternative viewpoints through weblogs and tagging. *Journal of Documentation 64(4),* 552-575.

Doctorow, C., Dornfest, F., Johnson, J. & Powers, S. (2002). *Essential blogging.* USA: O'reilly.

Dvorak, J. C. (2002). The blog phenomenon, *PC Magazine.* Retrieved March, 21, 2008 from:www.pcmag.com/article2/0,1759,81500,00.

Dye, J. (2006). Folksonomy: a game of high-tech (and hi stakes) tag: Should a robot dictate the terms of your search?. *Econtent 29(3)*, 38-44.

Eszter's Blog (2003). *Blog types.* Retrieved March, 2, 2008 from: http://campuscgi.princeton.edu/~eszter/weblog/archives/00000191.html.

Farkas, M. (2007). The bloggers among us. *Library Journal, 132(20)*, 40.

Farkas's Blog (2008). *Information wants to be free.* Retrieved September 22, 2008 from: http://meredith.wolfwater.com/wordpress/2008/09/08/how-i- got-started-blogging/.

Fichter, D. (2006). Internet applications for tagging and folksonomies. *Online 30(3)*, 43-45.

Golder, S. A. , & Huberman, B. A. (2006). Using patterns of collaborative systems. *Journal of Information Science 32(2)*, 198-208.

Gossett, J. L. & Byrne, S. (2002). Click here: A content analysis of Internet rape sites. *Gender and Society 16*, 689-709.

Graham, B. L. (2002). Why I weblog: a rumination on where the hell I'm going with this website. In J. Rodzvilla (ed.) *We've got blog: How weblogs are changing our culture.* Retrieved March, 21, 2008 from: www.bradlands.com/words/maybe/maybe02.html.

Haberman, J. (2005, March). *Analysis of the usage and value of Weblogs as a source of business news and information.* Retrieved March 20, 2008, from: http://blog. jackvinson.com/documents/Report Analysis of the usage and value of weblogs Julia Habe. Pdf.

Hall, A. & Davison, B. (2007). Social software as support in hybrid learning environments: The value of the blog as a tool for reflective learning and peer support. *Library and Information Science Research 29(2)*, 163-187.

Herring, S. C., Scheidt, L. A., Kouper, I. & Wright, E. (2006). A longitudinal content analysis of weblogs: 2003-2004. In M. Tremayne (Ed.), *Blogging, Citizenship, and the Future of Media* (3-20). London: Routledge.

Huffaker, D. A. , & Calvert, S. L. (2005). Gender, identity, and language use in teenage blogs. *Journal of Computer-Mediated Communication, 10* (2), Retrieved March, 21, 2008 from: http://jcmc. indiana.edu/vol10/issue12/huffaker.html.

Lasica, J. D. (2005). Blogging as a new form of journalism: Weblogs offer vital, creative outlet for alternative voices. In J. Rodzilla (ed.) *We've Got Blog: How Weblogs are Changing our Culture.* Retrieved March, 26, 2008 from: http://www.ojr.org/ ojr/workplace/1060217106.php.

Lee King's Blog (2008). *David Lee King.* Retrieved September 26, 2008 from: http://www.davidleeking.com/.

Marketingterms.com, *Blogs* (2007). Retrieved February, 20, 2008 from: http://www. marketingterms.com/dictionary/blog/.

Mathes, A. (2004). *Folksonomies - cooperative classification and communication through shared metadata*. Retrieved February, 20, 2008 from: http://www.adammathes.com/academic/computer-mediated-communication/folksonomies.html.

Mernit, S. (2003). *Blogger classifications: Some thoughts*. Retrieved March, 20, 2008 from: http://susanmernit.blogspot.com/archives/2003_10_05_susanmernit_archive.html#106554742635066352.

Merriam Webster Online, *Blogs* (2004). Retrieved February, 25, 2008 from: www.merriam-webster.com/dictionary/blogs.

Nardi, B. A., Schiano, D. J., Gumbrecht, M. & Swartz, L. (2004). Why we blog, *Communications of the ACM*, *47(12)*, 41-46.

Noruzi, A. (2006). Folksonomies: (Un)controlled vocabulary. *Knowledge Organization*, *33 (4)*, 199-203.

Noruzi, A. (2007). Folksonomies: Why do we need controlled vocabulary? *Webology 4(2),* Retrieved February, 24, 2008 from: http://www.webology.ir/2007/v4n2/editorial12.html.

Peterson, E. (2006). Beneath the metadata. *D-Lib Magazine*, *12*(11) Retrieved January, 27, 2008 from: http://www.dlib.org/dlib/november06/peterson/11peterson.html.

Pew/Internet American life project, *A portrait of the internet's storytellers* (2006). Retrieved January, 20, 2008 from: http://www.pewinternet.org Pew/Internet American life project, *New data on blogs and blogging* (2005). Retrieved January, 20, 2008 from: http://www.pewinternet.org.

Quintarelli, E. (2005). *Folksonomies: Power to the people*. Retrieved May, 20, 2008 from:http://www.iskoi.org/doc.folksonomies. htm.

Salant, A. (2006). *Evaluation of information on the web: New perspectives*. Retrieved September 26, 2008 from: http://portal.macam.ac.il/ArticlePage.aspx?id=1231&referer=/Default.aspx?.

Schaap, F. (2004). *Multimodal interactions and singular selves: Dutch weblogs and home pages in the context of everyday life*. Paper presented at AoIR 5.0, Brighton, 19-22 September.

Sinclair, J. & Cardew-Hall, M. (2008). The folksonomy tag cloud: When is it useful? *Journal of Information Science 34(1)*, 15-29.

Speller, E. (2007). Collaborative tagging, folksonomies, distributed cassification or ethnoclassification: A literature review. *Library Student Journal* February (2007). Retrieved January, 22, 2008 from: http://www.librarystudentjournal.org/ index.php/lsj/article/viewArticle/45/58.

Spieteri, L. F. (2006). The use of folksonomies in public library catalogues. *The Serials Librarian 51(2)*, 75-89.

Spieteri, L. F. (2007). The structure and form of folksonomy tags: The road to the public library catalog. *Information Technology and Libraries 26(3)*, 13-25.

Stephens, M. (2007). Web 2.0, Library 2.0, and the hyperlinked library. *Serials Review 33 (1)*, 253-256.

Thompson, G. (2003). Weblogs, warblogs, the public sphere, and bubbles. *Transformations*, 7. Retrieved January, 22, 2008 from: http://transformations.cqu.edu.au/journal/issue_2007/article_2002.

Turgeon, M. C. (2004). *10 reasons why blogging is good for you*. Retrieved January, 25, 2008 from: http://mcturgeon.com/blog/2004/11/24/10reasonstoblog/.

Vander Wal, T. (2006). *Understanding folksonomiey: Tagging that works.* Retrieved January, 29, 2008 from: http://s3.amazonows.com/2006presetations/docnstruct/Tagging_in_RW pdf.

Webopedia, *Blog* (2004). Retrieved January, 28, 2008 from: webopedia.com/TERM/b/blog. Html.

White, H. D. (2001). Computing a curriculum: Descriptor-based domain. Analysis for educators. *Information Processing and Management 37,* 91-117.

In: Internet Issues: Blogging, the Digital Divide and Digital... ISBN: 978-1-60741-824-5
Editor: Cassie M. Evans, pp. 95-116

Chapter 6

THE DATA DELUGE: OPPORTUNITIES FOR RESEARCH IN EDUCATIONAL DIGITAL LIBRARIES

Beijie Xu[1*], Mimi Recker[1] and Sherry Hsi[2]

[1]Department of Instructional Technology & Learning Sciences,
Utah State University, Logan, UT, USA
[2]Lawrence Hall of Science, Berkeley, CA, USA

ABSTRACT

Increasingly, educational digital libraries and associated services that support both formal and informal learning are available on the Web. As these collections are web-based, they can be engineered to capture fine-grained web usage data. However, because of their voluminous and detailed nature, rigorous analytical and interpretation strategies remain challenging. In response, *webmetrics* refers to the process of measuring, collecting, analyzing, and reporting web usage data. This chapter discusses the utility of webmetrics as a research method, and illustrates approaches for engineering and collecting web usage data in two digital library environments, the Instructional Architect and the Exploratorium's Learning Resources Collection. The chapter describes how using a set of common webmetrics such as visits, page views, and geographical location can help better understand the behaviors of end users. We also describe the application of *educational data mining (EDM)* to help discover meaningful patterns in usage data. Results include analyses from a pilot study that applies a clustering technique to teacher usage data from the Instructional Architect. The chapter concludes with a discussion of some of the current limitations and future possibilities of using web metrics to analyze and evaluate use and impact of educational digital libraries.

* Corresponding author: Email: beijie.xu@aggiemail.usu.edu

INTRODUCTION

Increasingly, educational digital libraries and associated services that support both formal and informal learning are available on the Web. The purpose of these sites is to provide teacher and learner access to high-quality learning resources in order to help improve both the effectiveness and efficiency of education (McArthur & Zia, 2008). In the U.S., the National Science Digital Library (NSDL.org) is a prominent example. This educational digital library offers a comprehensive collection of over 2.5 million high-quality, catalogued educational content and services to learners, educators, and academic policy makers (McArthur & Zia, 2008). The Exploratorium's Learning Resource Collection (nsdl.exploratorium.edu) is another example digital library and a focus of this chapter. It offers access to over 700 teacher-tested science activities and instructional resources inspired and created from the Exploratorium's exhibits, public program events, and teacher professional development programs.

As these collections are web-based, they can be engineered to capture fine-grained web usage data. The resulting massive and longitudinal datasets potentially contain important evidence about teacher and learner behaviors in complex online information ecologies (Computing Research Association, 2005; Borgman et al., 2008). However, because of their voluminous and detailed nature, rigorous analytical and interpretation strategies remain challenging. This challenge is sometimes referred to the 'data deluge' problem (Hey & Trefethen, 2003).

In response, *webmetrics* refers to the process of measuring, collecting, analyzing, and reporting web traffic through a web site (Chan, 2008; Khoo et al., 2008). In tandem, the field of *educational data mining (EDM)* offers methods for mining and discovering patterns in data automatically captured by educational tools (Muehlenbrock, 2005; Nickles, 2005; Romero & Ventura, 2007). These approaches typically do not assume conventional models or employ traditional statistical methods; instead they rely upon computational power to help reveal and visualize emerging patterns.

Engaging in rigorous and ongoing webmetrics research is a key component of a comprehensive evaluation strategy of digital library environments. First, as a baseline, it helps provide usage data, and thus another means for understanding teacher/learner activities that are increasingly happening outside of formal school hours. Second, it can help determine if the site is meeting end user needs. For example, are teachers/learners using the site as intended? Why or why not? Third, identifying unexpected usage can help designers learn from users, and identify new avenues for growth. Finally, the data provide fertile ground for applying emerging analytical techniques to complex educational data.

This chapter describes approaches for engineering educational digital libraries and services to capture, analyze, and mine web usage data. In doing so, we first review key webmetrics that are applicable to the study of web usage activity. We next present strategies for describing these large datasets. We then discuss an application of an emerging field of study, educational data mining, to help discover meaningful patterns in usage data.

To ground the discussion, we demonstrate the utility of these methods in the context of two digital library environments, the Instructional Architect (IA), and the Exploratorium's Learning Resources Collection (ELRC). We conclude with a discussion of important limitations when working with these kinds of datasets.

Description of Two Educational Digital Library Environments

The Instructional Architect

The Instructional Architect (IA) is an educational digital library service developed to support the instructional use of the National Science Digital Library (NSDL) and other online learning resources. With the IA, teachers are able to search, select, sequence, annotate and reuse online learning resources to create instructional web pages, called IA projects. IA breaks down the technology barriers and allows teachers with basic computer skills to create online projects and efficiently address their instructional needs (Recker et al., 2007).

Audience

The Instructional Architect targets two main audiences: teachers and their students. A teacher can create a free account, which provides exclusive access to his/her saved resources and projects. As part of the registration procedure, the teacher completes a profile indicating subjects and grades taught, teaching experience, and level of information literacy.

System description

After a teacher logs into the system, the IA offers two major usage modes: resources management and project management. In the resources management mode, teachers can search for and store links to NSDL resources within the IA context (see Figure 1), as well as provide a name and add non-NSDL resource links to their own collection too. In addition to an individual default resource folder, a user can create and name more folders for organizing saved resources.

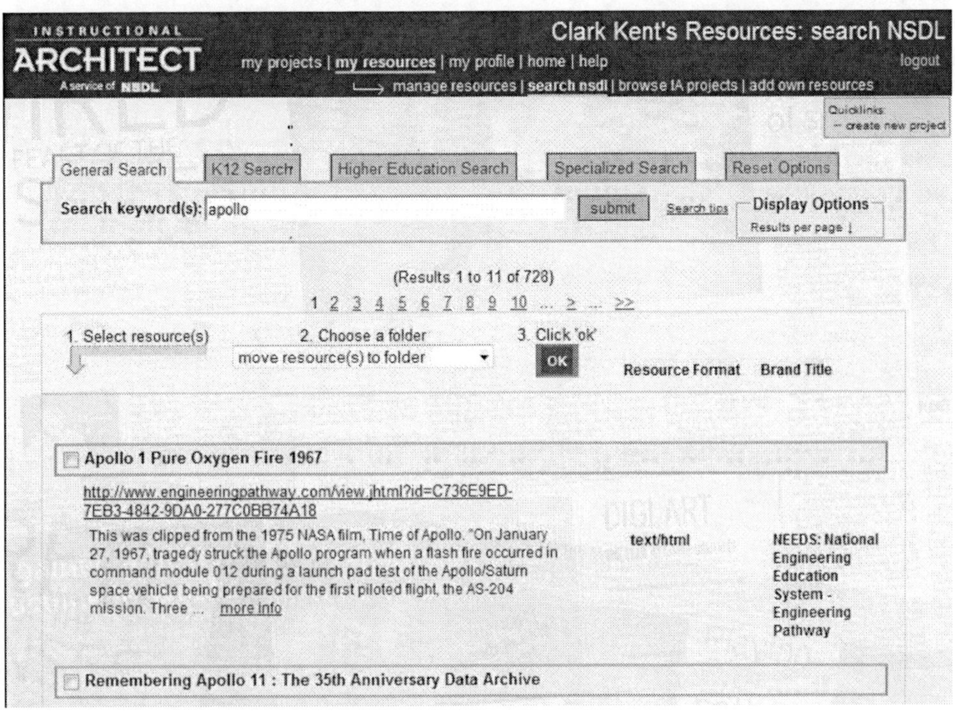

Figure 1. Searching for NSDL Resources inside the IA

The project management mode allows teachers to create an IA project web page and share it with the public or only with their own students. The teacher's resource collections are listed on the left, and can be added to the project with a single click (see Figure 2). JavaScript and HTML code is allowed, which means dynamic objects such as multimedia, blogs, and RSS can be included. Teachers can add basic project metadata such as subject area, grade level, and core curriculum standard, and these metadata are used to support project search and browse. A registered teacher can create a generic student account that is shared by all his/her students. With such an account, students have exclusive access to their teachers' private projects that are marked as "student-view only".

Web usage

From 2002 to June 2009, over 4,100 users have registered in Instructional Architect, more than 8,500 IA projects have been created, and 38,000 online resources have been added to the IA database. Since August 2006, IA public projects have been viewed over 560,000 times (see also Table 2).

The Exploratorium learning resources collection

The Exploratorium Learning Resources Collection (ELRC) (nsdl.exploratorium.edu) is a digital library of over 700 teacher-tested science activities and instructional resources inspired and created from the Exploratorium[1]'s exhibits, public program events, and teacher professional development programs.

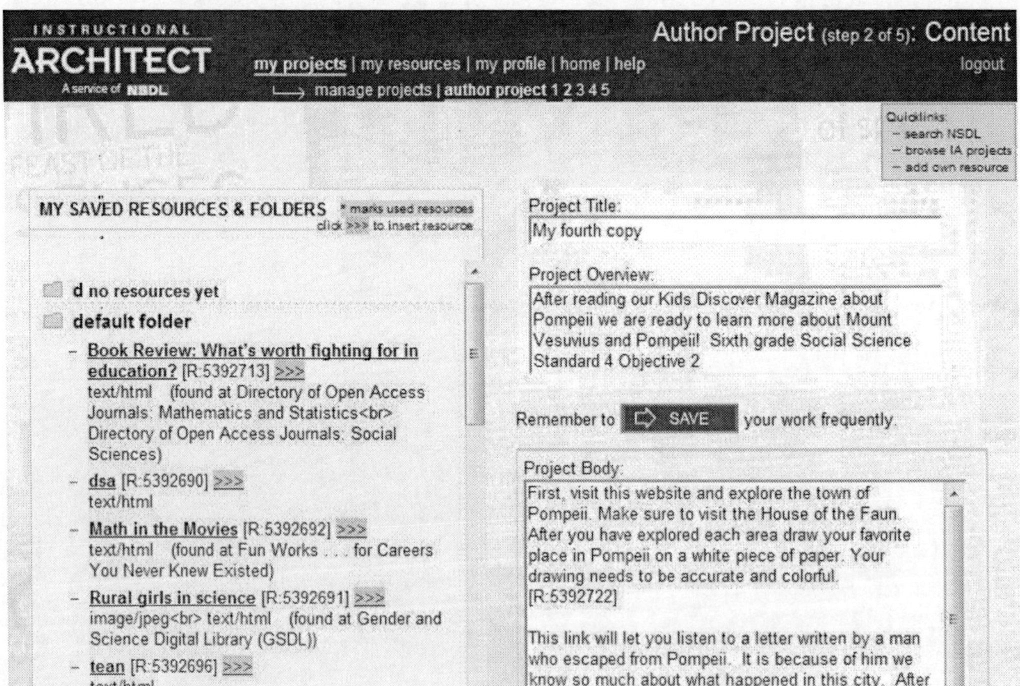

Figure 2. Project Creation Interface

[1] The Exploratorium is a hands-on museum of science, art, and human perception located in San Francisco, California.

The ELRC is designed for elementary and secondary school teachers as a primary audience, and informal educators as a secondary audience. Teachers can browse the collection by topic or conduct keyword searches. Search results provide a short description of the item, as well as related topics to explore (Figure 3). For each resource item found, a resource record is provided that includes a description of the resource as well as teaching tips where appropriate. Advanced search enables teachers to narrow a search by curricular area, grade level, and specific resource type (i.e., image, video, activity, article, web interactive, web exhibition, museum exhibit, and professional development resource.) The ELRC is available to use from the Web and no registration is required to use the application. Through interoperability, the ELRC items can be also found in other educational libraries including the National Science Digital Library (nsdl.org).

Web Usage

In 2008, the digital library ELRC was accessed by over 35,500 unique visitors. Since its launch in 2005, visitors from all 50 States across North America and over 170 countries have accessed the collection. The most frequently accessed resources include the website page from "Faultline" which explains the differences between P- and S waves, and a hands-on electric circuit activity called "Jitterbug."

Figure 3. Search results from the Exploratorium Learning Resources Collection

WEBMETRICS DATA SOURCES

More than a dozen commercial and open source Web Analytics tools are currently available. One approach, called page-tagging, uses third-party services that capture web traffic by embedding Javascript into web pages. Popular examples include Google Analytics (analytics.google.com) and Omniture (www.omniture.com).

Another approach relies on analyzing log files that are automatically generated by web servers. Examples include Summary (summary.net) and AWStats (awstats.sourceforge.net). For dynamically generated sites, such as the IA and the ELRC, a third approach relies on engineering the back-end database to record fine-grained user transactions for later analysis. Together, these tools can capture a complex array of data (see Table 1 for examples).

The IA has been collecting web server logs since 2002, and engineered to collect data with Google Analytics (GA) since early 2006. As a dynamic site, its database captures many low-level actions on a per-user basis. Finally, registration profiles are also collected when teachers create their free IA accounts. This includes grade level, subjects taught, teaching experience, and information literacy skills.

The Exploratorium's main website has been collecting basic webmetrics data since 1995 using the commercial tool *Summary* (Semper et al., 2000), and with Google Analytics since 2007. The ELRC collects several sources of data that include web traffic from GA, server log files, search terms, and a short online user survey. Because the Exploratorium was interested in broader impact, an embedded evaluation tool was also created to request zip codes of users who both lived in the U.S. and visited the site, providing evidence of use by people from other country.

Currently, there is little standardization across Webmetrics tools in defining and tracking user visits (Google Analytics Report, 2009). As such, caution must be taken when comparing them. For example, Figure 4 plots page views as reported by GA and the IA database over a 15-month period. Note that while the results can differ by as much as 12%, the correlation is very high (r=.98).

Table 1. Common webmetrics

Metric	Definition
Visits	Sequence of user transactions from a single IP within a certain time period (considered a more robust measure than 'hits')
Unique visitors	A measure of a site's true audience size
Page views	Number of times web pages are accessed during a single visit, including repeated viewing of the same page
Visitor origin	Country or region of visitor origin
Operating Sys/ Browsers	Operating system and browser used by visitors
Session length	Time spent by visitor during each single visit
Entry/Exit Pages	First and last pages accessed by visitors
Referrer	Last page the user visits before landing on the site
Bounce rate	The percentage of visitors who "bounce" away to a different site from the entrance page
Search keywords	The search terms that visitors enter into a search engine

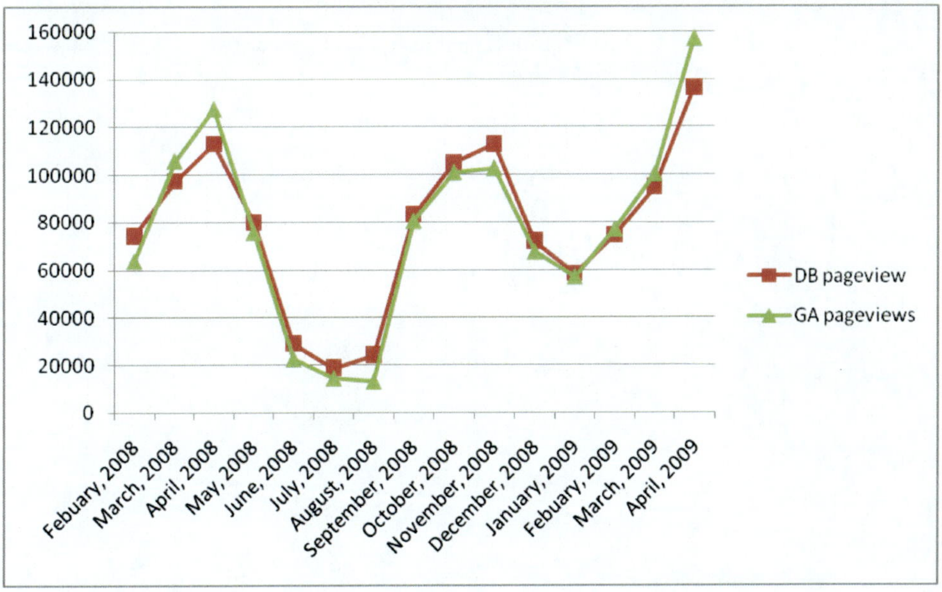

Figure 4. Page Views from Two Webmetrics Sources over a 15-month Period

WEBMETRICS DESCRIPTION

Working with large web usage datasets provides many challenges. This section describes two important considerations: data aggregation and tracking growth.

Data Aggregation and Presentation

Data aggregation and presentation strategies involve analyzing important metrics for a particular learning context on a regular basis. These strategies help address questions about users, their behaviors, their location, and the content they are accessing.

Users

First, it is important to note that most web usage datasets show an underlying zipf (or long-tail) distribution (Recker & Pitkow, 1996). Thus, a first step in examining data is to plot users against some activity measure (for example, number of logins). Figure 5 shows a plot of the number of users against the number of IA projects created. It displays the characteristic long-tail distribution, except in the middle. Noting such distributions means that some statistical measures (e.g., means, standard deviation) and tests (t-test) should be used with caution because the dataset might not meet underlying assumptions.

Google Analytics can also be a powerful tool in addressing questions about users, in that it can quickly report a variety of aggregated usage data over different time periods. For example, Figure 6 shows visits to the IA and ELRC over a 12-month period. Note that while both graphs show holiday ebbs, the IA's shows a dramatic drop-off during the summer reflecting its school-based usage.

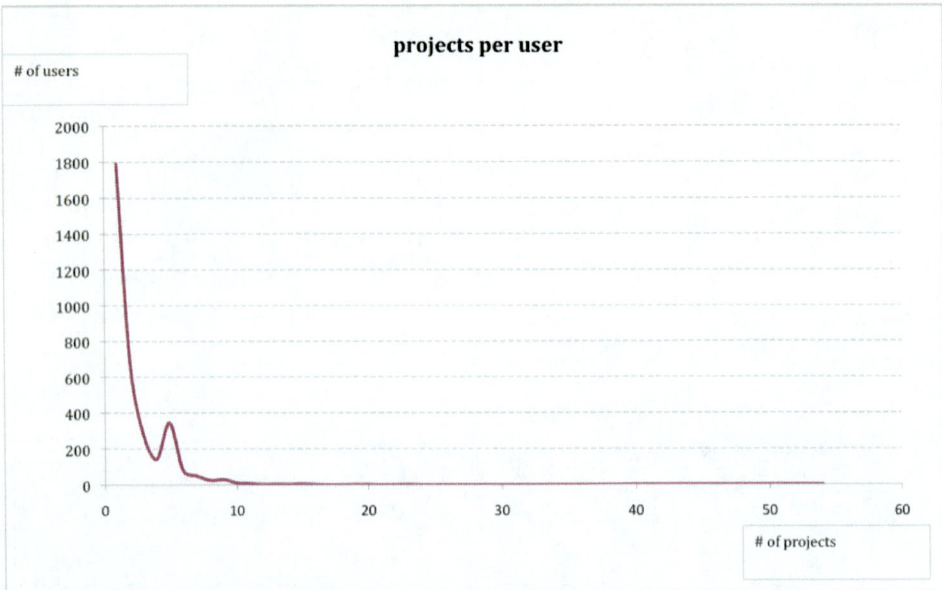

Figure 5. A Zipf (long-tail) distribution characterizes IA's usage

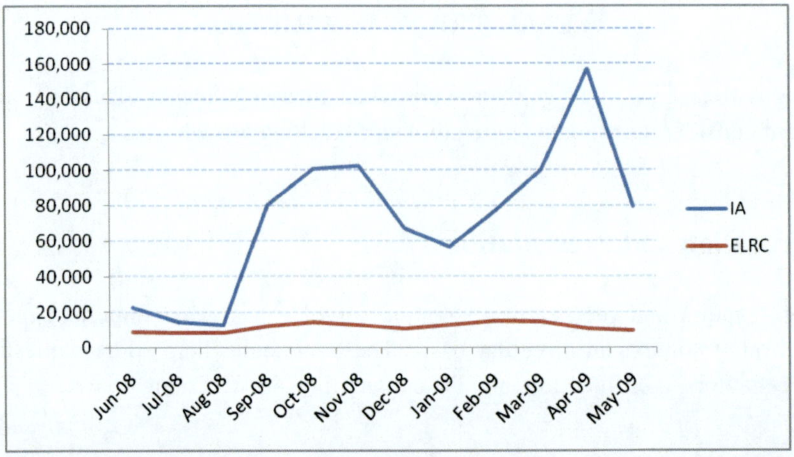

Figure 6. Number of visits, as generated by Google Analytics

In contrast, the ELRC aims to understand how many visitors were educators, and how many educators were first time users or loyal users of the web resource. By aggregating data and sorting data by IP addresses, and making an assumption based on *.edu* are educational users from schools and universities, one can infer what proportion of visitors are from educational institutions and therefore educators. The temporal pattern of higher traffic during the school year is also an indicator that formal educators are using the resource.

When analyzing the two sites' traffic source over a similar time period, the IA shows a much higher percentage of direct traffic than ELRC (57.45% vs. 27.39%), and lower bounce rate (33.93% vs. 54.89%). This suggests that more ELRC users are discovering the site, while the IA has more repeated users and purposeful usage.

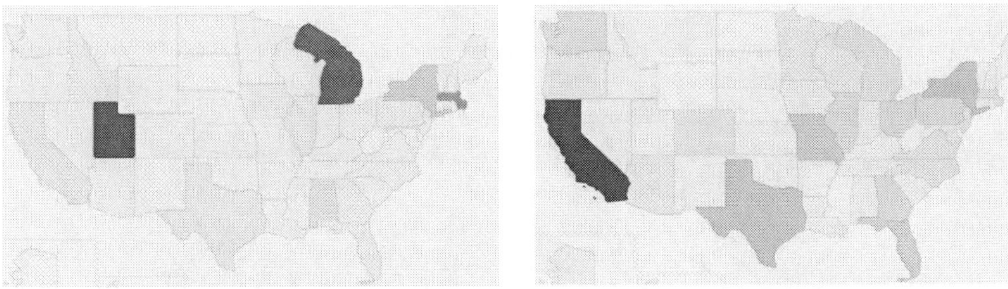

(a), Instructional Architect (b), ELRC

Figure 7a&b. U.S. Geo-segmentation generated by Google Analytics. Darker shade indicates more visits

Finally, analyzing *length of session* reveals interesting differences in patterns between the two sites. As an authoring site, 15.9% of IA sessions last more than 10 minutes. In contrast for the ELRC, a more 'shop-and-go' site, 6.9% of sessions last more than 10 minutes.

Location

Webmetrics can help determine the geographical location of site visitors, called *geo-segmentation*. Figure 7a&b shows visits to the IA and ELRC, respectively, over a 6-month period in the US. Darker shades indicate more visits. Given that the IA is based in Utah and has outreach activities in New York and Michigan, while the ELRC is California-based, the maps show that both groups are successful in local dissemination activities. The ELRC also shows more widespread U.S. visitors.

These maps can also be overlayed with other publically available geographical datasets, for example from the U.S. Census (factfinder.census.gov) and the National Center for Education Statistics (nces.ed.gov). These maps may reveal relationships between site usage and demographic or school characteristics.

Figures[2] 8a&b and 9a&b show IA and ELRC visitor traffic overlayed on two selected datasets (number of schools per district and median family income), as they most appeared to display a relationship with visitor traffic. The relationship with number of schools suggests that visitors are frequently school audiences. The relationship with family income level may be a function of resources (i.e., computing) available in the local schools and communities.

Content

An examination of popular content accessed on sites can reveal insights about users' intent. In terms the IA, 26% of the use represents user accesses to teacher-created IA projects. GA also shows that the ratio of student login paths to teacher login paths was nearly constant, at about 4:1. This suggests the site is used as intended: teachers create IA projects for their students to use.

For the ELRC, the top page is the search results page, which occurs four times as often as the browse page (the second highest page). One can infer that educators have a task in mind and are looking for specific resources rather than browsing the collection.

[2] For space reasons, only the continental 48 states are shown.

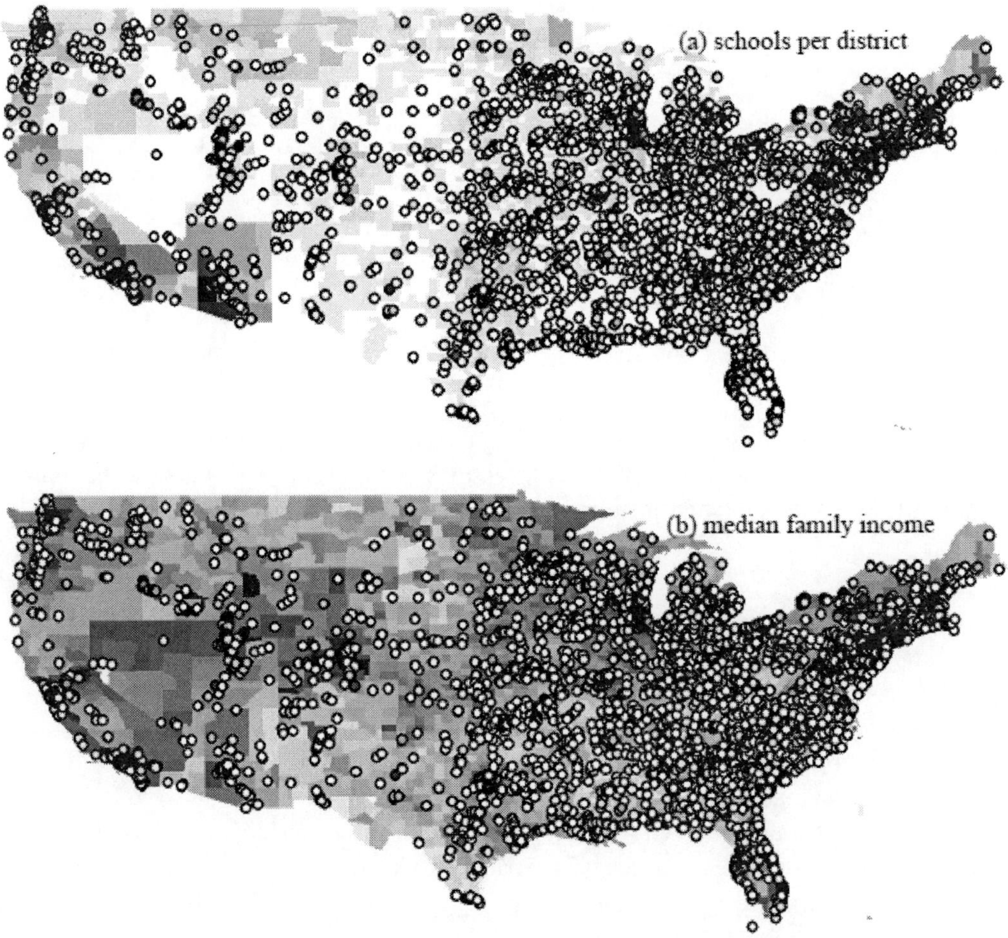

Figure 8a&b. U.S. map showing IA visits (darker dot is high visit frequency) overlayed with (a) number of schools per district, and (b) median family income (darker shade indicates higher value) over 1 year.

Table 2. Usage data growth (to June 2009)

	Since	Number	Growth %
Registered users	09/2004	4,119	36
Web resources used	01/2005	37,691	50
IA projects created	09/2003	8,513	50
Project views	08/2006	565,000	58

Tracking Site Growth

Tracking site growth addresses questions such as identifying usage growth, and how it compares to other sites. Here, it is important to avoid comparing raw visit counts, as sites are very different in their purposes and penetration, and may be at different phases of maturity. Instead, percentage growth for particular time periods can be more insightful. For example, Table 2 shows growth rate in IA usage in various categories. It shows the largest growth in the number of project views.

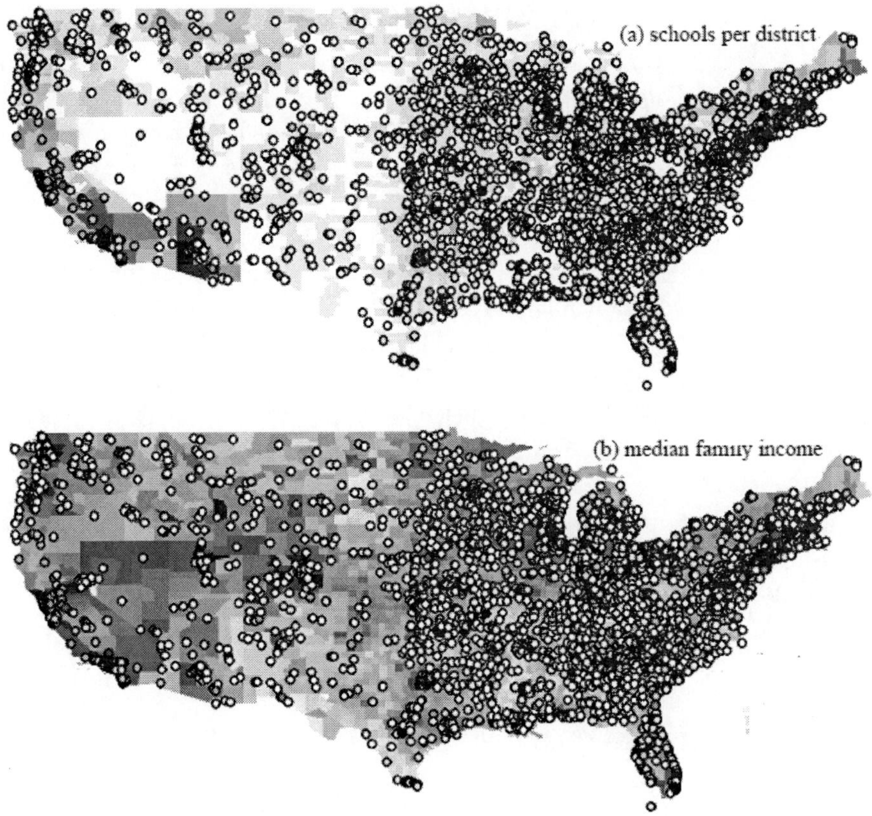

Figure 9a&b. U.S. map of ELRC visits (darker dot is high visit frequency) overlayed with (a) number of schools per district, and (b) median family income (darker shade indicates higher value) over 1.5 years.

Comparing growth for the same site in different years can also reveal the impact of a dissemination or marketing strategy. For example, in April 2009, teacher-created IA projects became available via the search portal at the National Science Digital Library (NSDL.org). To identify if this resulted in increased discovery, we compared traffic originating from the NSDL portal over a similar time period in 2008 and 2009. As shown in Figure 10, the year 2009 shows a marked increase, evidence of the effectiveness of this dissemination strategy.

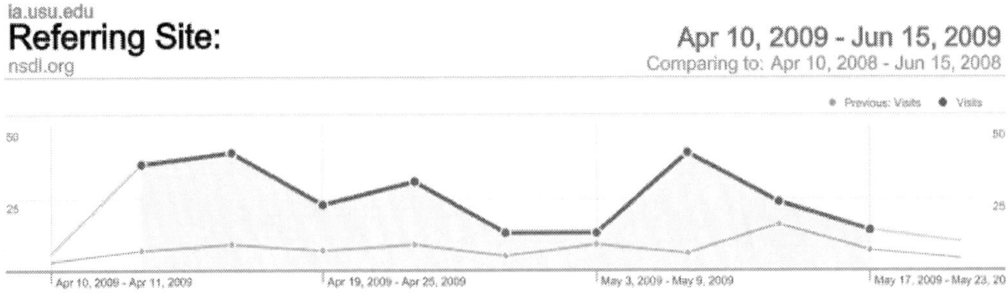

Figure 10. IA Traffic Referrals from the NSDL months before (green) and after (blue) dissemination strategy.

APPLICATION OF EDUCATIONAL DATA MINING: PATTERN DISCOVERY

There is an growing interest in data mining (DM) and the evaluation of web-based educational systems, making educational data mining (EDM) a rising and promising research field (Romero & Ventura, 2007). Data mining is the discovery and extraction of implicit knowledge from one or more large collection of data (Pahl & Donnellan, 2002; Romero & Ventura, 2007), and when the context is the Web, it is sometimes explicitly termed web mining (Chen & Chau, 2004; Cooley, Mobasher, & Srivastava, 1997). Educational data mining applies the data mining process to an educational dataset, making inquiries about the site's impact, usage, its users and the users' behaviors. Web-based educational applications are able to record the users' fine-grained behavior in real-time in a log file or a database, providing a huge amount of data for the educators to analyze and hence better understand the user profile (Romero & Ventura, 2007).

This section uses the Instructional Architect as an example to illustrate the application of data mining to understand an educational digital library's users and their usage patterns. We begin by describing the dataset, then review our general EDM processes, and conclude with results from a pilot study.

The Instructional Architect Dataset

The primary dataset for the application of educational data mining algorithms is the IA relational database. In addition to the objects and data related to IA functionality, the database also stores user traces: a table called *saved_projects* stores every project's past versions, a window to examine how the teachers develop and revamp their IA work; any hit on an IA resource or an IA project is recorded in *a tracking_hits* table, which keeps the IP address, user ID, timestamp, session ID, referrer page, target object (either IA resource or IA project); a *tracking_page_hits* table provides even more detailed footprints – it contains similar fields as the *tracking_hits* table, and in addition to requests to IA resources and projects, it records almost every URL (a PHP page or link to an external resource) users have clicked.

Knowledge Discovery in Database

Educational data mining by and large follows the standard Knowledge Discovery and Data Mining (KDD) process: 1) data cleaning and integration, selection and transformation, 2) applying data mining algorithms, 3) evaluation and presentation (Han & Kamber, 2002; Witten & Frank, 2005). Sometimes, the first two phases (data cleaning and integration, selection and transformation) are combined and called data preprocessing (Cooley, et al., 1997; Romero & Ventura, 2007).

This section describes major issues and challenges in each KDD phase documented in the literature.

Phase I – Data Preprocessing

Raw data are far from being ready to be input into a mining algorithm, because there might be missing entries, irrelevant information, or the need to integrate data from different resources before using them. Thus, the first step of knowledge discovery – data preprocessing – is very critical to ensure that the data are in a suitable shape and could produce valid results. It is estimated that data preprocessing comprises the majority (90%) of the work in knowledge discovery. Yet, it is more than a tedious necessity, because all the decisions made in data preprocessing phase can deeply influence the results of the actual data mining. Data preprocessing in general may contain more than one of the tasks listed below.

Data Cleaning and Path Completion

Data cleaning, the most intensive step in data preprocessing, is to remove noise and inconsistent data from the data source (Han & Kamber, 2002). Search engine spiders (Weischedel & Huizingh, 2006) constantly crawl IA pages, and spam programs have attacked the IA in the past, leading to a large number of spam accounts, projects, and visits scattered in different tables of the database. In addition, some teachers attempted to create projects, but stopped in the middle, leaving numerous unfinished work (e.g., without title or without content) in the system. Thus, all the data contributed from noise made by spiders and humans need to be removed from the database.

Data Integration

Data integration entails the combination of data from multiple autonomous and heterogeneous sources (Halevy, Rajaraman, & Ordille, 2006; Han & Kamber, 2002; Romero & Ventura, 2007). It is another central step in data preprocessing for knowledge discovery (Kriegel, et al., 2006). Data mining algorithms generally require a homogeneous dataset – data originating from one single source. However, sometimes information from a single source is insufficient for data mining, and there is a need to refer to different sources to get a more comprehensive picture of the topic at hand, especially in real-world situations, such as large enterprises who own a multitude of data sources, large-scale scientific projects that collect data produced independently by different researchers, or cooperation among government agencies (Halevy, et al., 2006). Entries of different data sources, however, may have different semantics, different scales and formats, and varying degrees of noise – all pose considerable challenges to data combination.

Data selection

Usually not all the information obtained from the raw data is necessary for data mining. For example, IA teachers are asked to select the background color and theme logo for their projects, information not significantly reflecting the quality and usage of the projects, and therefore should be discarded from the set of prepared variables. As such, a set of features – feature vector – is selected to represent a user's profile and their relevant online activity.

Google Analytics is useful in this regard. It presents a series of reports about the visitors, traffic source, and navigation path, which can provide a lead for the direction of further exploration. Table 3 illustrates how the GA data influence the data selection process.

Data transformation

Data transformation is the process of transforming raw data into forms appropriate for mining. Han & Kamber (2002) lists several commonly used data transformation methods: smoothing (e.g., binning, regression, and clustering), aggregation (e.g., converting daily sales data to monthly total amount), generalization (replacing lower-level or primitive data with higher-level concepts), normalization, and attribute construction (adding new attributes from the given attributes).

These methods are utilized to reduce the complexity of the IA dataset. Table 4 presents an example of the data transformation adopted in the pilot study to be discussed later.

Table 3. Google Analytics (GA) and Data Selection

GA facts	Questions for further exploration	Relevant variables
40% visitors have only visited the IA site once, while 8% have visited it more than 200 times.	- Can different behaviors be identified from users of different visit frequency? - What are the typical behaviors of loyal users?	- The duration between visits - The number of times a user has visited the IA. - The number of projects a user has created. - The number of IA resources a user has collected.
40% people bounce away after viewing one page and the percentage of visitors decrease as they go deeper into the site; however, 4% have more than 20 pages per visit.	- Is the depth of visit a descriptive attribute for categorizing users?	- Depth of visit
Particular IA projects become the top landing pages.	- Is the projects' popularity a descriptive attribute for categorizing users?	- A project's number of visits
GA has observed some popular navigational path.	- How to generalize those navigational paths into meaningful activities, in order to characterize visitors?	- Page requests
Site overlay tells that a guest user usually visits three places after entering the site: my resources, create new projects, and browse IA projects.	- Is a users' frequently visited page(s) a descriptive attribute for categorizing users?	- Page requests

Table 4. Data Transformation

Raw data	Transformed data	Method
Timestamps	Session length	Generalization Attribute construction
Number of times in various occasions	Project / resource / visit stickiness	Smoothing Generalization
Project types	The percentage of each type of projects	Attribute construction
URL	Browsing activity	Generalization
Number of page views per visits	Depth of visit	Generalization Attribute construction

Phase II – Applying Data Mining Algorithms

After the raw data are cleaned, transformed and important variables are selected, some data mining algorithms will be applied to extract the hidden patterns from the processed data. Researchers in education and information science have applied almost all data mining algorithms to discover useful information from educational datasets. This pilot study, described below, relied on a particular type of data mining method -- clustering -- in analyzing IA users and their usage patterns from different perspectives.

Clustering

Clustering is an unsupervised learning model (Han & Kamber, 2002). That is, there is neither a predefined number of clusters nor pre-labeled instances. Clustering algorithms normally group data based on two measures: the similarity between the data objects within the same cluster (minimal intra-cluster distance), and the dissimilarity between the data objects of different clusters (maximal inter-cluster distance).

In the IA context, teacher users are the most complicated group of visitors; they exhibit different navigational behaviors, and make their individual decisions on how to use web resources and their own projects. It is very likely that more than one type of teachers exist in this complex IA environment. Since there are no predefined teacher sub-groups, clustering algorithms should suit the situation – to discover typical IA teacher groups.

Phase III – Interpretation and Post-processing

There is no golden standard for evaluating the 'goodness' of mining results, and even the same dataset could render different results after going through different preprocessing procedure and web mining algorithms. In addition, the interpretation is highly problem dependent. Just as neither p-values nor effect sizes make sense unless contextualized and accompanied with appropriate explanations, the same applies to data mining results. The discovered patterns would not be very useful unless there are mechanisms and tools to help analysts better understand them (Cooley, et al., 1997). Although statistical analysis software (e.g., SPSS, LatentGold) and web analytics tools (e.g., Google Analytics) have the ability to visually display results, researchers are ultimately responsible for the interpretation and presentation of discovered patterns.

Pilot Study: Clustering IA Users

In this section, we present a pilot study using a small sample of Instructional Architect usage data. The purpose of this study was to detect and describe user groups based on their varying online behaviors.

Data Preprocessing

User selection

The small-scale pilot study included users who registered between January and June 2009. From this set, data from one-time users and users who never created any projects were excluded. The data from the remaining 468 registered users (out a total of 560 registered users during that period) were included in the pilot study.

Feature selection

We identified one category of process measures (Nickles, 2005), *navigational behavior*, and two categories of product measures, *resource usage* and *project usage,* to quantify and aggregate teacher users' online behavior. Several usage variables were used to define each category (see Table 5).

Some values were easily calculated, such as the percentage of each type of projects. Some were aggregated from numerous footprints, such as session length, and depth of visit. Finally, because users have been using the IA for different amount of time, it is unfair to compare the number of visits without considering how long they have been registered. As such, some variables were constructed by comparing them to other users' performance, such as resource stickiness, visit stickiness and project stickiness. The latter are described next.

First, we calculated every user's number of visits occurring in the *i*th month after registration, denoted by v_{ki} (k = 1, …, 468). Second, we calculated the average number of visits occurring in the *i*th month after lining up all users' registration date. That is, average(v_i) = $\overline{v_{ki}}$, where $v_{ki} \neq 0$. Then, we calculated the average number of visits occurring in the first *i*th months after registration. Sum(i) = $\sum_{j=1}^{i} \overline{v_{kj}}$, where $v_{kj} \neq 0$. Finally, each user's visit stickiness is calculated by comparing the total number of visits with sum(i) where i is the number of months since registration. Thus, stickiness(k) = v_k / sum(i).

Clustering

This pilot study used Latent Class Analysis (LCA) (Magidson & Vermunt, 2002) to classify registered users (teachers) into groups. LCA is a model-based cluster analysis technique in that a statistical model (a mixture of probability distributions) is postulated for the population based on a set of sample data. LCA offers several advantages over traditional clustering approaches such as K-means: 1) for each data point, it assigns a probability to the cluster membership, instead of relying on the distances to biased cluster means; 2) it provides various diagnostics such as common statistics, L2, and p-value to determine the number of clusters and the significance of variables' effect; 3) it accepts variables of mixed types without the need to standardize them; and 4) it allows for the inclusion of demographics and other exogenous variables either as active or inactive factors (Magidson & Vermunt, 2002, 2004; Vermunt & Magidson, 2002).

Initially, all variables were entered into the latent class analysis as indicators, using three to six cluster models to group teachers. Some variables had an R^2 less than 0.1, meaning those user features did not contribute much to the explanation of group membership. Moreover, such phenomena were consistent across different models. Thus, variables with less discriminative power were removed from all models one by one, leaving only eight features in the final analysis (see the last column of Table 5).

Table 5. Feature Space

Categories	Features	Data type	Description	Cluster Analysis
Resource Usage	Add own resource	Categorical	Add a non-NSDL resource	Removed
	Add IA projects as resources	Categorical	Put an IA project into the resource collection	Removed
	Resource stickiness	Continuous	# of resources collected by this user divided by average # of resources by all users after N months of use	Included
Project usage	Project stickiness	Continuous	# of projects by this user divided by average # of projects by all users after N months of use	Included
	Percent of public projects	Continuous	The percentages of public projects among all this user's projects.	Included
	Percent of student-view projects	Continuous	The percentages of student-view projects among all this user's projects.	Included
	Percent of private projects	Continuous	The percentages of private projects in this user's projects.	Included
	Percent of copied projects	Continuous	The percentages of copied projects in all this user's projects.	Included
Navigation	Browse other projects	Categorical	Has this user browsed other people's projects?	Included
	Visit stickiness	Continuous	# of visits by this user divided by average # of visits by all users after N months of use	Included
	Session length	Continuous	The average time spent on the IA website per visit	Removed
	Depth of visit	Continuous	The average # of pages visited per session	Removed

When continuous indicators are used, the cluster module can be designed to range from the most unrestricted to the most restricted models. With an unrestricted model, each cluster may have its own variance and a full covariance matrix; though flexible, it results in a large number of parameters to be estimated, which increases as the indicators and the number of clusters k increase. On the other hand, if we assume all clusters share the same variance and all covariances equal zero (locally independent), we get the most restrictive model, requiring less parameters but relying on an unrealistic assumption.

In order to set up the most parsimonious probability model, we allowed for class dependent variance and set the off-diagonal elements of the covariance matrix to be zero, and set the number of clusters from $k = 3$ to 6. In this round, all indicators achieved statistical significance ($p < 0.05$) and a reasonable R^2 (> 0.1). But when $k = 6$, one of the cluster's size was less than 1%, indicating that this model overestimated the number of clusters. On the other hand, when $k = 3$, the model failed to provide clear distinctions between the users of different clusters. Finally, we set $k = 5$, because it produced a much lower BIC value (a Bayesian information criterion to measure model fitness) than when $k = 4$.

In the 5-class model, we observed several large bivariate residuals (BVR), suggesting that the model fell somewhat short of explaining the association between variables. Thus, we included local dependence between pairs of indicators with large BVR one by one, until all

BVR values were less than 10. With such specification, we achieved much lower BIC value, and accepted such setting as the final model.

Results and Interpretation

Clustering results

Table 6 shows the final clustering results. The values under each cluster are the mean scores for the corresponding indicators.

Table 6. LCA Clustering Results

		Cluster1	Cluster2	Cluster3	Cluster4	Cluster5
Cluster Size		0.29	0.24	0.21	0.16	0.10
Indicators	Range					
Percent_student_projects	0 - 1	0.96	0	0.34	1	0.44
Percent_private_projects	0 - 1	0	1	0.29	0	0.32
Percent_copy_projects	0 - 1	0	0.15	0.30	0	0.32
Percent_public_projects	0 - 1	0.96	0	0.40	0	0.43
Visit stickiness	0.22 – 8.93	1.16	0.62	0.99	0.71	2.98
Project stickiness	0.15 – 8.06	1.99	0.20	0.66	0.32	2.12
Resource stickiness	0 - 29.91	1.06	0.39	0.77	0.57	4.22
Browse others	0 - 1	0.44	0.44	0.20	0.59	0.21

Note. k = 5. Class dependent variance and local dependence between the following pairs of indicators: resource stickiness and project stickiness, percent_public_project and percent_private_project, percent_public_project and percent_student_project. Browse others is the only dichotomous variable. 0 means a teacher has never browsed another user's project, and 1 means browsed at least once.

Interpretation

We examined each variable's distribution with the following interpretations about the characteristics of each user group, ordered by cluster size.

Cluster 1: Goal-oriented, willing-to-share

This group of users tends to collect many web resources and is very productive (high project stickiness). They never create dummy projects, and once they have an instructional need, they will turn it into a public IA project or make it available to their students. Though they may or may not browse other people's projects, they don't adapt others' work, perhaps because they prefer a customized lesson plan to meet students' learning objective.

Cluster 2: Inactive, less motivated

This group of users does not visit the IA very often. Moreover, members have never contributed to the IA community because they only create a few private projects for practice or just for fun.

Cluster 3: Lukewarm performers

This group of users shows loading on every indictor, yet all stickiness indicators measure at lower levels. Members in this cluster have adapted other teachers' projects. Though they have created several projects, they make at least one project public or available to their students. They appreciate and utilize every function provided by the IA service, but are less enthusiastic than Cluster 5.

Cluster 4: Goal-oriented, unwilling-to-share, less devoted

Like users in Cluster 1, people in this group also create IA projects with specific instructional objectives in mind. However, they only make projects available to their own students, unwilling to share with the general public. As infrequent users, they only collect a few resources and create a few projects. Probably they use IA for one time to fulfill their instructional needs but seldom come back.

Cluster 5: Productive, active, adapters

Members of this group like to browse other teachers' projects and make a copy for themselves. They are more active and productive than group 3 in terms of visits, projects created, and resources collected. Though extremely active, they are not as focused as Cluster 1 and Cluster 4.

Summary

As noted above, the IA is intended to increase the utility of online resources for teachers. Teachers are expected to collect and contextualize resources, and share their instructional web pages with students and colleagues. Cluster 1 seems to be the group that best meets this objective, as members are active consumers, designers, and contributors to this teacher network (Recker, 2006). However, the fact that they seldom adapt others' work makes them less productive than Cluster 5. Ideally, we expect teachers to possess the strength of both groups to become functional and efficient IA users.

CONCLUSION

This chapter has reviewed approaches for analyzing web usage data, webmetrics and educational data mining, in the context of two learning environment, as well as presented a pilot study with teacher usage data to demonstrate clustering using Latent Class Analysis. Our work identified diverse user behaviors, and as well as clusters of effective users.

In the future, we plan to add another layer of data transformation by converting all variables to a 0 ~ 1 range. We also plan to extend the current work to a large-scale study by including all users to analyze their longitudinal behaviors. The present pilot study only considered eight indicators, and ignored vast amount of user data, such as click-stream data. Educational data mining is an emerging field, where researchers bring together knowledge from different fields to answer educational research questions. In the long term, our research is not necessarily limited to one methodology and a single algorithm, as we plan to utilize other data mining approach (e.g., association rule mining) and to combing results with more traditional evaluation data (e.g., user surveys) to better illuminate the behaviors and motivations of IA users.

In conclusion, we note that while these approaches have utility, they do come with several caveats. The implementation of webmetrics tools as well as analysis of the resulting data is time consuming. Effective webmetrics analyses require adequate resources, integrated with a comprehensive research and evaluation strategy. A second caveat is limitation of the webmetrics, which are a measure of user behavior not of knowledge or attitude. Consequently, inferences about users' intentions cannot be made solely from web metrics. To help with interpretation issues, usage data should be triangulated with other data sources, such as data from usability studies, field studies, surveys, focus groups, and interviews. Finally, because collection of web usage data can be done without the full knowledge of the user, it poses a privacy threat. Informed consent needs to be designed into the online environment to allow users to opt-in or out of research participation

Despite the current struggle, the field of web metrics and educational data mining is slowly making progress towards standardizing terms and measures. As online environments continue to generate a data deluge of massive and longitudinal datasets, this presents multiple opportunities to explore this rich territory.

ACKNOWLEDGMENTS

This material is based upon work supported by the National Science Foundation under Grants No. 840745 & 0840738, and Utah State University. Any opinions, findings, and conclusions or recommendations expressed in this material are those of the author(s) and do not necessarily reflect the views of the National Science Foundation. We thank the members of the IA research group, especially Bart Palmer.

REFERENCES

Borgman, C. L., Abelson, H., Dirks, L., Johnson, R., Koedinger, K. R., Linn, M. C., et al. (2008). *Fostering learning in the networked world: The cyberlearning opportunity and challenge.* Report of the NSF Task Force on Cyberlearning. Retrieved from http://www.nsf.gov/pubs/2008/nsf08204/nsf08204.pdf.

Chan, S. (2008). Towards new metrics of success for on-line museum projects. In: J. Trant, & D. Bearman, (Eds.), *Proceedings of the Museums and the Web 2008.* Toronto, Canada: Archives & Museum Informatics.

Chen, H. & Chau, M. (2004). Web mining: Machine learning for web applications. In C. Blaise (Eds.), *Annual review of information science and technology, 38* (289-329). Medford, NJ: Information Today, Inc.

Computing Research Association (2005). *Cyberinfrastructure for education and learning for the future: A vision and research agenda.* Retrieved from Computing Research Association, CRA publication website: http://www.cra.org/reports/cyberinfrastructure.pdf

Cooley, R., Mobasher, B. & Srivastava, J. (1997). Web mining: Information and pattern discovery on the World Web Web. *Proceedings of the 9th IEEE International Conference on Tools with Artificial Intelligence, Newport Beach, CA,* 558-567.

Google Analytics Report (2009). Why does Google Analytics report different values than some other web analytics solutions? [Online troubleshooting reports]. Retrieved June 08, 2009, from http://www.google.com/support/googleanalytics/bin/answer.py?hl= en& answer=55614.

Halevy, A., Rajaraman, A. & Ordille, J. (2006). Data integration: The teenage years. *Proceedings of the 32nd International Conference on Very Large Data Bases, Seoul, Korea,* 9-16.

Han, J. & Kamber, M. (2002). *Data mining: Concepts and techniques* (2nd ed.). San Francisco, CA: Morgan Kaufmann Publishers.

Hey, T. & Trefethen, A. (2003). The data deluge: An e-Science perspective. In F. Berman, G. Fox & T. Hey (Eds.), *Grid computing: Making the global infrastructure a reality* (809-824), Wiley.

Khoo, M., Recker, M., Pagano, J., Palmer, B., Washington, A. & Donahue, R. A. (2008). Using webmectrics to analyze digital libraries, *ACM/IEEE-CS Joint Conference on Digital Libraries*. Pittsburgh, PA.

Kriegel, H. P., Borgwardt, K., M, Kröger, P., Pryakhin, A., Schubert, M. & Zimek, A. (2006). Future trends in data mining. *Data Mining and Knowledge Discovery, 15(1)*, 87-97.

Magidson, J. & Vermunt, J. (2002). Latent class models for clustering: A comparison with K-means. *Canadian Journal of Marketing Research, 20*, 37-44.

Magidson, J. & Vermunt, J., K. (2004). Latent class models. In D. Kaplan (Ed.), *The Sage Handbook of Quantitative Methodology for the Social Sciences* (pp. 175-198). Thousand Oakes: CA: Sage Publications.

McArthur, D. J. & Zia, L. L. (2008). From NSDL 1.0 to NSDL 2.0: Towards a Comprehensive Cyberinfrastructure for Teaching and Learning. *Proceedings of the JCDL'08, Pittsburg, Pennsylvania*, 66-69.

Muehlenbrock, M. (2005). Automatic action analysis in an interactive learning environment. *Proceedings of the workshop on Usage Analysis in Learning Systems at AIED-2005, Amsterdam, The Netherlands.*

Nickles III, G. M. (2005). Identifying measures of student behavior from interaction with a course management system. *Journal of Educational Technology Systems, 34(1)*, 111-126.

Pahl, C. & Donnellan, D. (2002). Data mining technology for the evaluation of web-based teaching and learning systems. *Proceedings of E-Learn 2002 World Conference on E-Learning in Corporate, Government, Healthcare, & Higher Education, Montreal, Quebec, Canada,* 747-752.

Recker, M. & Pitkow, J. (1996). Predicting document access in large, multimedia repositories. *ACM Transactions on Computer-Human Interaction , 3(4)*, 352-375.

Recker, M. (2006). Perspectives on Teachers as Digital Library Users: Consumers, Contributors, and Designers. *D-Lib Magazine, 12(9).*

Recker, M., Walker, A., Giersch, S., Mao, X., Halioris, S., Palmer, B., et al. (2007). A study of teachers' use of online learning resources to design classroom activities. *New Review of Hypermedia and Multimedia, 13(2)*, 117-134.

Romero, C. & Ventura, S. (2007). Educational data mining: A survey from 1995 to 2005. *Expert Systems with Applications, 33(1)*, 135-146.

Semper, R., Wanner, N., Jackson, R. & Bazley, M. (2000, April). *Who's Out There? A pilot user study of educational Web resources by the Science Learning Network (SLN).* Paper presented at the Museums & the Web 2000 Conference, Minneapolis, MN.

Vermunt, J., K. & Magidson, J. (2002). Latent class cluster analysis. In J. Hagenaars & A. McCutcheon (Eds.), *Applied latent class analysis* (89-106). New York, NY: Cambridge University Press.

Weischedel, B. & Huizingh, E. K. (2006). Website optimization with web metrics: A case study. *Proceedings of the 8th International Conference on Electronic Commerce, Fredericton, New Brunswick, Cananda,* 463-470.

Witten, I. H. & Frank, E. (2005). *Data mining: Practical machine learning tools and techniques* (2nd ed.). San Francisco, CA: Morgan Kaufmann.

In: Internet Issues: Blogging, the Digital Divide and Digital... ISBN: 978-1-60741-824-5
Editor: Cassie M. Evans, pp. 117-129

Chapter 7

THE ROLE OF FAMILY ON INCREASING PARTICIPATION AND BRIDGING THE DIGITAL DIVIDE: EVIDENCE FROM AN E-HEALTH STUDY

Sally Lindsay[1*], Simon Smith[2] and Paul Bellaby[3]

[1] Bloorview Research Institute, Bloorview Kids Rehab, Toronto, and Dalla Lana School of Public Health, University of Toronto, Ontario Canada

[2] Institute of Communication Studies, University of Leeds, Leeds UK

[3] Institute for Social, Cultural and Policy Research, University of Salford, Salford UK

ABSTRACT

Despite the rapid increase in the diffusion of information and communication technology (ICT) the significance of older people's connection with their families on their willingness to use the Internet remains unclear. The aim of this chapter is to examine how family members helped to bridge the digital divide by influencing older people's use of the Internet. Our sample includes older people (n=108) from a deprived area of Greater Manchester, UK who have heart problems. Thus, they have much to gain from 'virtual mobility' because they may be less likely to have home Internet access or to visit public Internet facilities. On the other hand, since they are likely to have fewer pre-existing online social ties, their initial inclination may be lower. We draw on the qualitative comments from the focus groups we conducted (n=30) alongside a 9-month randomized control trial in addition to comments posted on our project portal. Our study adds to the literature by examining the advantages of *home* computer and Internet access for those previously without them, while most studies focus on *public* use or those who already own computers. Further, by focusing on a sample that encounters multiple exclusions (e.g., deprived financially, poor health, older) we can develop a better understanding of the complex nature of the digital divide. Our findings suggest that grandchildren played an important part in the uptake and continued use of their computers for several people. Although some people reported that family relations had

* Corresponding author: E-mail: slindsay@bloorview.ca

some negative influences on their Internet use, most had a positive experience and even developed new common interests with children, grandchildren, siblings and their spouses. We draw on Bourdieu's concept of habitus to understand the role that family had in shaping Internet use.

INTRODUCTION

Although the majority of households now own a computer with Internet access, some people still do not (National Statistics 2005). This persisting digital divide between the haves and the have nots mirrors other social divides based on social class, income, education, place of residence and age (Norris 2001; OxIS 2005). Digital exclusion by social class and age are well understood and efforts to overcome digital divides typically address lack of economic capital and practical training. Yet, navigating through the Internet successfully also requires reflexivity: discrimination among sources, self-confidence and a balance of trust in and wariness about virtual encounters (that is, cultural and social capital) (Lindsay et al. 2007; Lindsay and Smith 2008). The 'habitus' (i.e., habitual pre-dispositions and cultural pathways) of relatively deprived and older people may disable them from taking the benefits that the e-society seems to hold out.

It is particularly worthwhile to examine older people because they are often at risk of social exclusion due to poverty, disability or poor health. The Internet can be useful in this regard because it can provide opportunities to improve quality of life that would otherwise be unavailable because of cost or limited mobility (Olphert et al. 2005). The e-society is expected to offset social exclusion and empower individuals who might otherwise be unable to manage their own lives (Lindsay et al. 2007; 2008a; 2009). Access to the Internet might facilitate exchanges between people in similar situations, who would not meet face-to-face. It might also provide access to information that people need, but could not get by other means (Lindsay et al. 2007; 2008a; 2008b; 2009).

Our study addresses several gaps in the digital divide literature. Here we offer *home* Internet access and computers to those without, while most studies focus on *public* use. Far less is known about how the Internet is used in the domestic environment among older people (Cushman & Klecun 2005a; 2005b; OxIS 2005). Finally, by focusing on a sample that encounters multiple exclusions (e.g., deprived financially, poor health, elderly) we can develop a better understanding of the complex nature of the digital divide.

UNDERSTANDING THE DIGITAL DIVIDE

Despite recent developments in public and community provision of ICT some 40% of the UK population do not use the Internet (Selwyn 2003). This is particularly the case for older people. Selwyn (2003) found that 85% of those older than 60 did not use the Internet. Rates of non-use are highest among those with fewer educational qualifications and those whose most accessible Internet location was either someone else's house or a public site.

According to Dutton, the lower rate of Internet use among older people is an example where "choice is shaped by habit or a perception that the Internet is not important"

(University of Oxford 2006). A national survey for Age Concern (2002) found that 66% of those aged over-55 who were non-users said that they would never use the Internet. This is compounded by the fact that older people have access to significantly less peer or family reinforcement in relation to using the Internet and are also less likely to have encountered computers at work or in education (Fox 2004; Olphert et al. 2005). This places older people in a situation where they have to make an active choice to use the Internet.

People tend to adopt technology in ways that fit in with their lives and do not take away time from other activities (Gershuny 2002). If these conditions are not met, for example if the mode of access is unfamiliar or the technology's usefulness is difficult to contextualise, the barriers to adoption are greater. For example, Melenhorst et al. (2001) found that a lack of perceived benefits inhibited email use in the communication of older people (more so than perceptions of cost). If, however, the benefits were contextualised in relation to people's own social networks and everyday settings, older adults may become more motivated to make the investment (of time and effort) in learning how to use a tool like email (Melenhorst et al. 2001). Similarly, Tatnall and Lepa (2003) found that the Internet was 'translated' as a task-specific, purposeful and simplified technology, whose meaning was strongly determined by the current activities and inter-personal networks of each individual. This assumes that a person's own social network is sufficiently technologically-enabled to sustain communication by email, which may not always be the case in the most impoverished communities. Cushman and Klecun (2005a) also discovered that first-time e-literacy students found it difficult to imagine what they might do with ICTs and had a deficient knowledge about the technologies. This lack of context made it difficult for them to develop plans for using ICTs. In motivating older adults to adopt technology it is particularly vital to be able to convey the benefits so that the entry thresholds for technology adoption are lowered (Melenhorst et al. 2001; Selwyn et al. 2003).

Many of the above observations apply equally to people living with health problems where just 40% of Americans of all ages who are living with a disability have Internet access (Pew Internet and American Life Project 2001). Although older people who use the Internet often do so to find health information (Fox 2004; Olphert et al. 2005), this does not imply that older people as a whole appreciate the usefulness of the Internet for managing their health. Evidence from inner city Manchester shows an inverse relationship between interest in online health resources and reliance upon their GP (Rogers and Mead 2003). Non-users had more trust in their GP, downgrading the value of lay knowledge, and had little 'positive outcome expectancy' in relation to their own information-seeking activities whether online or offline (Rogers & Mead 2003). Invoking concepts of self-efficacy and role legitimacy, Rogers and Mead (2003) claim the failure of a free Internet clinic as evidence that access is insufficient to encourage people to explore the Internet for health resources. It may be that older people with a long-term condition are more likely to have low self-efficacy and a traditional notion of role legitimacy. Finnish researchers found that self-assessed health was a significant factor in predicting Internet use, with healthier respondents among older people more likely to be users (Vuori & Holmlund-Rytkonen 2005).

Although governments have invested heavily in public Internet access provision in an effort to increase participation in the information society evaluations show that the impact on social exclusion has been slight (Hall Aitken 2003). Estimates based on an evaluation of UK Online centres for DfES suggest that at best around a quarter of current non-users of ICT can be 'included' by a policy of public provision alone (Hall Aitken 2003). In all EU countries the users of public Internet access points (PIAPs) are disproportionately younger, better educated and are likely to be multiple-location Internet users rather than people who use public facilities alone (European Opinion Research Group 2001). PIAPs can be a valuable supplementary resource within communities especially if there is appropriate support to learners or if the facility is already an important site for community-building, social networking or group formation (Smith et al. 2004). In these rare circumstances community associations and neighbourhood groups can act effectively as 'informal Internet support groups' for socially excluded people (Foley 2004). PIAPs are not, however, an effective safety net for the digitally excluded. Their value is particularly dubious for some older people, given that 37% of people aged 65 and over spend 'all or nearly all' of their leisure time at home, compared to 17% of all UK adults. Those with a disability, those aged 75 and older, and those living alone are even more likely to do so (Ofcom 2006). On the one hand there is a huge knowledge gap about opportunities in the community. It is doubtful whether better publicity would make much difference, since these places themselves are often not perceived by older people as comfortable or secure compared to their own home (Olphert et al. 2005). This provides some insight as to why only 10% of non-users aged over 55 agreed that better learning facilities would encourage them to take up IT (Age Concern 2002).

There is little research on older people's use and non-use of digital technologies that relates directly to their usage of and attitudes towards the Internet. What information we have is piecemeal, frequently contradictory and mainly quantitative rather than qualitative (Olphert et al. 2005; Vuori & Holmlund-Rytkonen 2005). We do, however, know with a fair degree of certainty that mere access to the Internet is not in itself sufficient to increase the use of ICTs by many categories of older people (Lindsay and Smith 2008; Lindsay et al. 2008a, 2008b; 2009). What remains largely unknown is the role of family in helping to bridge the digital divide amongst older people.

There are, indeed, good reasons for believing that family can act as a catalyst for technology adoption. Family ties are usually the first form of social intermediation to which an individual has recourse. This is significant because evidence is beginning to emerge that social intermediation can be an extremely important enabler of access to online resources. For example, a cross-European survey found that 42% of e-government service users also assisted family or friends (Millard 2006), and these technology disseminators typically acted "at family and community levels," often within poor neighbourhoods (DG INFSOC 2007: 24). Crang et al (2006), in a study examining patterns of Internet use in areas of different levels of affluence, found that in the less well-off neighbourhoods, where everyday urban lives remain reliant on cash economies, face-to-face services and physical travel, ICT use tended to develop as a collective, collaborative activity (whereas it was more individualised in the better-off areas). These findings underscore the importance of taking into consideration the forms of social intermediation – notably family ties – which are available to an individual from a disadvantaged background weighing up the pros and cons of technology use. Intermediation is particularly important if use is defined broadly to include situations in which people make use of online services thanks to intermediaries such as family members and other advocates (Smith 2009), although here we are concerned primarily with tactual physical use of the technology.

The domestication of new technology involves a two-way process where consumers change the meaning and influence of technologies while at the same time influencing family interactions (Katz & Rice 2002). Technological innovations are seen as a major source of social change within the social systems of families. Home Internet access may facilitate family communication by offering a new avenue for family members to communicate, potentially creating new functions and specializations (Mesch 2006). Changes in functions involve changing the nature of family relationships. Here we aim to explore the role of family in influencing older people's use of the Internet, and specifically whether family members can help to change habitus.

THEORISING THE INFLUENCE OF HABITUS ON INTERNET USE

In order to gain a deeper understanding of the role of family in bridging the digital divide we draw on the concept of 'habitus.' Immediate material circumstances have some bearing on enabling and diffusing changes in both Internet use and health-related behaviour, but probably less so than what Bourdieu (1984) calls *habitus*—the impact of cumulative material affluence/deprivation on 'embodied' social and cultural capital. This implies an embedded disposition sometimes overlaid by more ephemeral attributes that in the long-term shapes the ways individuals respond to the world around them. Physical and mental constitution is part of habitus and both are typically a legacy of early upbringing. The central issue of our research study was how and to what extent, through facilitated learning on the Internet, habitus may be changed among deprived older people resulting in a new lifestyle and outlook that would enable the disease to be tackled more effectively.

The concept of habitus suggests a static world in which, for instance, unequal health outcomes and exclusion from the benefits of an information society are passed on generation by generation. That is not altogether unrealistic, but it neglects the dynamic nature of class relations: the incremental, occasionally revolutionary, shifts that have occurred in societies in which land and mobile means of production are owned or controlled by classes that exploit the labour of the dispossessed. It also neglects the extent to which innovations in technology and ideas are produced by or diffused within newer generations.

Bourdieu's (1971) concepts of 'cultural and social capital, habitus and framing' suggest that something passive (i.e., not being 'tuned in') is the explanation for not using the Internet. Here we explore the idea that self-management of heart disease through the Internet is perceived as a *risk* that should be avoided.

When free computers and Internet access meets with lack of economic capital and/or lack of social capital, it tends to fall on stony ground. Similarly, where there is not cultural and/or social capital to take advantage of material opportunities, such as being given a PC and Internet access, the opportunity may not be taken (Lindsay and Smith 2008).

METHODS

This chapter draws on a larger study (see Lindsay et al 2008b; 2009) with a sample (n =108) of men and women aged 50-74 from GPs' coronary heart disease registries in Salford. Our sample was drawn from Salford in Greater Manchester, UK because nearly half of the electoral wards are in the top 10 per cent of the most multiply deprived wards in England, including an increased risk for heart disease (Salford PCT, 2005). The over-riding aim of this nine-month trial[1] was to test whether community based, facilitated access to an Internet health portal could improve the capacity of older people with heart disease to manage their own heart conditions. The sample was divided at random into an experimental and a control group of equal sizes. All participants were given new computers and a one-year broadband subscription; however, only the experimental group received training and access to the "Hearts of Salford" project portal. A technician installed the computers in the participants' homes and also assisted with any difficulties that arose. Weekly drop-in sessions and phone-in support was also available to both groups although we anticipated that the experimental group would be better informed about the drop-in sessions because they were promoted via the portal.

Hearts of Salford (http://www.heartsofsalford.net/)[2] was established as an online space where the experimental group could interact in one of five dedicated closed groups, with facilitation from researchers. The portal social architecture was initially constrained by the terms of the research design, in the sense that mixing between successively recruited groups was prevented until each had had the benefit of six months facilitated access. This was construed as a period of socialization and familiarization with the online environment, which took place within a safe private space for 10 people who had met one another in person at an introductory meeting and training session.

Questionnaires, focus groups, and Internet log data were collected from May 2006 to September 2007. Here we draw on the qualitative data that reveals the impact that these new computers had on the participant's family ties.

RESULTS

The results suggest that family members, especially grandchildren, helped to shape the habitus of older people in the uptake of the Internet. Although there were some instances of family members causing a negative experience, most reported a positive experience overall with having home access to the Internet.

Factors Influencing Take-Up and Continued Use of Technology

Several men and women reported being motivated to learn how to use a computer because their children and grandchildren were using them:

> *As far as computers are concerned every single one of my grandchildren can use them. So can my children. But I can't use the damn things because I've never had time to sit there and work it out and lately I've been too frightened to get it in case I won't get the other jobs done before I get on it because I am totally fascinated with the idea.* (#74, married female)

---

[1] Ethical approval for which was obtained from the Salford University Research Ethics Board and the local NHS research ethics committee.

[2] The website that exists today at this URL is a re-designed version of the original project portal, which is now managed by a community group formed from among the project participants. The original portal used an educational software platform called Moodle.

> *I've not got the foggiest idea about computers. I've only watched the grandchildren on theirs and they are fantastic but it's something I've always wanted to get into.* (#71, married male)

Grandchildren not only inspired their uptake of computers but they also helped to teach them. This was especially helpful when they got stuck in some application and did not know what to do next.

> *Sometimes quite by accident I come across something and I don't know how I did it. I'll have to get my grandson down to spend a day with me showing me how to do things he is 10 years and is a whiz. It's when you keep trying and not getting anywhere you lose interest that is when you could do with someone to show you what you are doing wrong. I haven't even tried the printer yet or the speakers. I have the same problem with the mobile phone, all I wanted it for was when I'm out and wanted a taxi but people will keep saying why don't you use the text? When I get things that are modern technology they seem to sense I get a mental block. Clever these computers, aren't they?* (#44, divorced female)

While some participants were inspired by their grandchildren others felt intimidated that they knew more than they did. A few men share their experience: "*Kids...Does it not intimidate you? It intimidates me*" (#41, married male). Another man similarly claimed:

> *You've all got grandkids and you see them there on that computer 10, 12, 14 years of age and they can type faster than you can talk and it is really jarring you off. You say how do you do this, oh it's just like that granddad.* (#64, married male).

Many participants described how they used their computer to help combat loneliness and boredom since many of these men and women lived on their own. One man captured it best: "*It has solved a problem of boredom. I can always find something interesting, informative and challenging to do*" (#46, divorced male). For others, adopting technology coincided with critical points of change in their own lives. For example, one woman shares her story:

> *I never used a computer till after my husband died. We bought him a computer because he needed something he could do. I looked at this computer after he died and thought well I don't even know how to turn it on. Seen him on it everyday researching into everything. I can play games on it I can write a letter on it now but it's taken years.* (#61, widowed female)

Those who had fewer people in their domestic space seemed to "seek out" technology in the hopes that they could engage more with people, while those who had closer family ties tended to have more of a 'push' factor where they were inspired or cajoled to learn by other family members. This suggests that family members (or lack thereof) conditioned the habitus of this older sample in ways that influenced their decisions to take up the Internet or not.

Impact of Technology on Family Life

Although these men and women found it difficult at first to overcome their fear of technology, with the help of family and friends they were able to engage better with their new computers. In fact, one participant commented how his computer, which was once viewed as something new in the household, is now viewed as commonplace. He said, "in regards to *the family where you have your computer, let's face it computers these days will eventually be part of the furniture. You have your fridge, freezer, microwave and now your computer*" (#43, single male).

Several participants described the benefits of increased connections with family members, not only through the Internet (via email or Internet phone call), but also around the computer itself in which family members would help participants to learn how to use it. For example, one person describes how he discovered the possibilities of phoning over the Internet: "*there is a program that might be of interest to the whole family, which is quite good. It's where you can use your computer as a telephone; it's free to other people anywhere in the world*" (#26, married male). The idea of saving money and still being able to connect with family was particularly appealing. Others were pleased that having broadband freed up the phone line so that they would not miss out on phone calls: "*It frees up the household phone line so that I can pursue links for as long as I wish without blocking phone calls to and from family and friends*" (#25, married male).

Other benefits of having a computer in the house included performing routine family-oriented tasks with greater ease. One woman describes her experience:

> *My son's family and my eldest grandson think this computer is brilliant. My grandson showed me how to go on websites and how to book travel and show tickets. It's so easy. We got car insurance sorted for our son. So all in all it's a great asset to our house!* (#56, divorced female)

Having a new computer in the household helped couples to pursue interests in a new way:

> *My husband thinks it is great that I can access queries I have about my health and how this has helped me. He also enjoys bringing up Manchester United and reading the news about the team. My grandchildren have helped us put our holiday photos into the computer and print them off.* (#52, married female)

Although there were several positive examples of the impact of these computers helping to strengthen family ties, there were also examples of it creating tensions between family members. One participant described how her son had tried to teach her on the computer but he became very frustrated and impatient and stormed off. One participant gives her example:

> *My lad is trying to teach me, he'll tell me that's how you do it and I'm looking at the screen while he's using his fingers I've not got a clue what he's teaching. So I've not learnt a thing. You've got to be patient. You should know it!* (#65, divorced female)

Another woman said that her computer is often a point of tension between her and her daughter (who still lives at home). She said that her daughter tries to control the computer and often puts a password-protected entry on it when they are in an argument to "get back" at her. This was a situation that negatively influenced her ability to use the Internet.

Meanwhile, several others report how they get frustrated when their grandchildren come by to use their computer and often 'mess things up.' For instance, one man said that he kept having pop-ups after his grandkids went on his computer:

> *There may be something wrong with my computer system because there are 3 or 4 little things that come up at the bottom all the time, trying to scan my computer. It's because the kids keep going on these games, and that's where they are trying to get in from. And I get warnings all the time flagging up.* (#54, married male)

Participants often found such problems very distressing as they are just beginners on the computer and like to have the desktop appear the same way each time. Another example of the computer having a negative impact came from one man who, although he was very happy with his computer claimed that his wife was upset with the amount of time he spends on it:

> *I look forward to going on the PC and the Internet. My world is now more open by pressing the click of the mouse. I am happy in the knowledge that I can use the Internet anytime but my wife does not agree. She thinks that I have another woman in the other room.* (#24, married male)

Indeed, ICT influenced family ties in both positive and negative ways. Their home Internet access introduced a social change and influenced habitus within their family network, which had both positive and negative implications. It created a new point of interest while allowing for easier communication among family members, which helped to strengthen family ties. Evidence of intergenerational conflict was also found especially around teaching older parents / grandparents how to use their computer.

Benefits of ICT in Supporting and Increasing Social Interactions within the Family

One key way that ICT was able to support and increase social interactions within the family was the ability to save money through the Internet. One man describes his experience:

> *I'm enjoying the learning challenge and have been able to help various members of my family in a number of ways. It has made me feel more useful and productive. I have been able to save money, gas and electric bills, shopping etc. This is a big help when on a low fixed income and frees up money for those extras and just generally eases the continual worry of 'making ends meet.' It has made me think I may eventually be able to generate an income doing book keeping and have been looking at online courses.* (#43, single male)

In this particular case, the computer helped to draw family ties closer together as they were able to save money and feel better about themselves. Another man gives a similar example:

> *It can have practical benefits. I have saved all my families at least 21% on their fuel bills. Helped my son and his family get a great deal on a holiday. Put minds at rest by getting complicated information from the Probate office.* (#46, divorced male)

Those who live on their own also benefited from increased connections with family:

> *I live on my own but I now email my nephew and a couple of friends and my buddies. My brother seems impressed by what I can do asking me to type out applications for golf club memberships and I've made some downloaded CDs for friends. I told my brother to have a go but he just won't get involved. My other brother has a computer but I think it's mainly his wife that uses it as he has had no formal training.* (#42, single male)

Another man living on his own describes how his computer became a focal point for his family:

> *My son brought his paperwork round today and asked me to see if I could find him a deal on gas and electric for his shop. He could probably have done it quicker for himself but it was nice to be asked. A few weeks ago he turned up with a 17 inch TFT monitor, which gives me a lot more room. My 2 daughters were round together last week. The youngest just helped herself to the computer to print off pages of detailed statements for the family's mobiles. I didn't know you could do that. My daughter-in-law got me three packs of paper. Last Saturday I had to promise my 2 youngest grandkids that I would be expert with the 'paint' application by next Saturday when they come again. It's like it's always been there but I'm slowly finding my way round it. It's a new point of conversation.* (#18, divorced male)

This particular man may have perceived a loss of status in the eyes of his family when he went from paid employment to dependence on benefits; however the services that he is now able to provide for family members seems to have restored some of his self-esteem. In this case habitus created an opportunity structure in which it was not difficult to see practical applications for the computer introduced into his house.

CONCLUSIONS

The introduction of technology into the household can influence family social dynamics. This project explored the significance of home Internet access for aging persons and their families. Our findings suggest that grandchildren played an important part in shaping older people's habitus in the uptake and continued use of their computers for several people. The introduction of these computers into the participants' homes had several positive impacts on their family ties such as a new common interest with children, grandchildren, siblings and spouse. Our findings are similar to recent studies that have found that the Internet enhances contact with family (Katz and Rice 2002; Kennedy and Wellman 2007; Mesch 2006). Here we found that the Internet can help improve family communication and provide opportunities for grandparents and grandchildren as well as parents and adult children to collaborate (Mesch 2006; Orleans & Laney 2000). Our findings are similar to the PEW Internet study (2001) which found that children act as Internet pioneers in households with respect to parents. Here we found that this is extended to grandparents, at least in close families.

There were some negative aspects that were also noted where family members tried to teach participants how to use their computer and ended up getting impatient with them. Some people were frustrated when another family member went on their computer and 'messed things up.' This is similar to other studies where Internet access in the household negatively altered family communication patterns (Mesch 2003; Watt & White 1999). We found that conflicts arose even though these older people were often not living with the family members who were using their computer or providing help.

Our results suggest that those who had fewer people in their domestic setting (single, or recently widowed) seemed to 'seek out' technology in the hopes that they could engage more with people, while those who had closer family ties tended to have more of a 'push' factor where they were inspired to learn by other family members (mainly grandkids). Some used their computers to connect to others outside of their household while others saw it as a way to engage with their grandchildren in their own social space.

Interaction with family members may reinforce existing habitus because family can provide "bonding social capital" whereas it is often "bridging capital" and new ties that may change it. Our findings here suggest that when people encountered new technology within a familiar environment, in the company of close family ties, they were able to adopt the Internet, without having to step outside their habitus. In this case, because the younger generations of their family had already absorbed ICT into the everyday surroundings of their lives, habitus therefore became a factor promoting Internet use by making it seem less risky. The participants seemed to conform to patterns of behaviour they observed among their children and grandchildren, rather than embarking on something new and unknown.

The role of habitus on people who live alone and do not have frequent contact with family appears to be weaker. Thus, they may have been more willing to explore risk (e.g., in seeking to form new social ties) and more responsive than those with closer family ties to the direct effects of the intervention. This research project provided some new resources that encouraged participants to be involved in something new that they otherwise might not have: the computer itself and access to the Internet (economic capital), the brief training we gave them (cultural capital) and perhaps even contact with and encouragement from the project team (social capital). Since Bourdieu (1971; 1984) tells us that it is easier to break habitus (to reposition yourself within the field) the more economic, cultural or social capital you have access to, each of these factors may have been important as they were intended to be according to the project design.

In summary, ICT helped to identify the internal dynamics of family life, which shaped the adoption and continued use of their computers. Family members seemed to play a role in influencing older people's habitus in deciding to use the Internet. Future research should explore in more depth the ways in which single older people use the Internet to communicate with their family and friends.

References

Age Concern (2002). *IT, the Internet and Older People*, Available at: http://www.ageconcern.org.uk/AgeConcern/5A158C9903E44DE8A1E8205448075B5F.asp.

Bourdieu, P. (1971). In (ed) Young. *Knowledge and Control: New Directions for the Sociology of Education*, London, Collier-Macmillan.

Bourdieu, P. (1984). *Distinction: A Social Critique of the Judgment of Taste*, Cambridge, Mass, Harvard University Press.

Crang, M., Crosbie, T. & Graham, S. (2006). Variable Geometries of Connection: Digital Divides and the Uses of Information Technology. *Urban Studies*, *43*, 2551-70.

Cushman, M. & Klecun, E. (2005a). *How (can) non-users perceive usefulness: bringing in the digitally excluded.* Penceil Paper 7, London School of Economics.

Cushman, M. & Klecun, E. (2005b). *Non-users of computers in south London.* Penceil Paper 6, London School of Economics.

DG INFSOC (European Commission Directorate General Information Society and Media) (2007). *Measuring progress in e-Inclusion.* Riga Dashboard 2007, Available at: http://ec.europa.eu/information_society/activities/einclusion/docs/i2010_initiative/rigadashboard.pdf.

Foley, P. (2004). Does the Internet help to overcome social exclusion? *Electronic Journal of e-Government,* no.2, paper 7.

Fox, S. (2004). *Older Americans and the Internet.* Pew Internet and American Life project, research report. March. Online: www.pewinternet.org.

Gershuny, J. (2002). Mass media, leisure and home IT: a panel time-diary approach. *IT and Society, 11(1)*, 53-66.

Hall Aitken Wyatt, J., Allison, S., Donoghue, D., Horton, P. & Kearney, K. (2003). *Evaluation of CMF funded UK online centres,* London, DfES Research Report No.502.

Katz, J. & Rice, R. (2002). *Social Consequences of Internet Use.* MIT Press, Cambridge.

Kennedy, T. & Wellman, B. (2007). The Networked Household. *Information, Communication & Society, 10(5)*, 645-670.

Lindsay, S., Smith, S., Bell F. & Bellaby, P. (2007). *Tackling the digital divide: exploring the impact of ICT on managing heart conditions.* Journal of Information, Communication & Society, *10(1)*, 95-114.

Lindsay, S. (2008). How and why the motivation and skill to self-manage coronary heart disease are socially unequal. *Research in the Sociology of Health Care, 26*, 17-39.

Lindsay, S. & Smith, S. (2008). The role of social capital in research participation. In: K., Gupta, G., Svendsen, P., Maiti (Editors) *Reinventing Trust: Discourses on Social Capital.* (Chapter 13) New Delhi: Atlantic Books.

Lindsay, S., Smith, S. & Bellaby, P. (2008a). Can informal e-learning bridge the digital divide? *Social Policy and Society, 7(3)*, 1-12.

Lindsay S., Bellaby, P., Smith, S. & Baker, R. (2008b). *Enabling healthy choices: Is information and communication technology the highway to health improvement?* Health: An Interdisciplinary Journal in the Study of Health and Illness, *12(3), 313-331.*

Lindsay, S., Smith, S., Baker, R. & Bellaby, P. (2009). The health impact of an online heart disease support group: A comparison of moderated vs. unmoderated support. *Health Education Research, 24(4)*, 646-654.

Melenhorst, A., Rogers, W. & Caylor, E. (2001). The use of communication Technologies by older adults: Exploring the benefits from the user's perspective. *Proceedings of the Human Factors and Ergonomics Society*, 45th Annual Meeting.

Mesch, G. (2003). The family and the internet. *Social Science Quarterly, 84*, 1038-1049.

Mesch, G. (2006). Family characteristics and intergenerational conflicts over the internet. *Information, Communication & Society, 9(4)*, 473-495.

Millard, J. (2006). *eGovernance and eParticipation: lessons from Europe in promoting inclusion and empowerment.* Paper presented at UNDESA workshop, Budapest.

National Statistics. (2005). *Consumer Durables.* National Statistics.

Norris, P. (2001). *Digital divide: civic engagement, information poverty and the Internet worldwide.* Cambridge. Cambridge University Press.

National Statistics. (2005). *Consumer Durables.* National Statistics. Ofcom (2006). *Media Literacy Audit: Report on media literacy amongst older people.* Available at: http://www.ofcom.org.uk/advice/media_literacy/medlitpub/medlitpubrss/older/ (accessed 14/6/09).

Olphert, C., Damodaran L. & May, A. (2005). *Towards digital inclusion—engaging older people in the digital world.* Accessible Design in the Digital World Conference.

Orleans, M. & Laney, M. (2000). Early adolescent social networks and computer use. *Social Science Computer Review, 18*, 56-72.

Oxford Internet Survey (Oxis). (2005). *Report the Internet in Britain.* University of Oxford: Oxford.

Pew Internet and American Life Project. (2001). *The online health care revolution: how the web helps Americans take better care of themselves.* Online: http://www.pewinternet.org

Phang, C., Sutanto, J., Kankanhalli, A., Li, Y., Tan, B. & Teo, H. (2006). Senior citizens' acceptance of information systems: A study in the context of e-government services. *IEEE Transactions on Engineering Management, 53(4)*, 555-569.

Rogers, A. & Mead, N. (2003). More than technology and access. *Health and Social Care in the Community, 12(2)*, 102-10.

Salford Primary Care Trust. (2005). *Improving the health of Salford.* Salford NHS PCT.

Selwyn, N. (2003). Apart from technology: Understanding people's non-use of information and communication technologies in everyday life. *Technology in Society, 25*, 99-116.

Selwyn, N., Gorard, S., Furlong J. & Madden, L. (2003). Older adults' use of information and communications technology in everyday life, *Ageing & Society, 23*, 561-582.

Smith, S., Maguiness, H. & Slocock, B. (2004). *The West Johnstone Digital Inclusion Project.* University of Paisley.

Smith, S. (2009). From inclusive spaces to inclusionary texts: how eparticipation can help overcome social exclusion. In Ferro, E., Dwivedi, Y., Gil-Garcia, J., Williams, M. (eds). *Handbook of Research on Overcoming Digital Divide: Constructing an equitable and competitive information society.* Hershey, PA: IGI Global.

Tatnall, A. & Lepa, J. (2003). The internet, e-commerce and older people: an actor-network approach to researching reasons for adoption and use. *Journal of Logistics and information Management, 16(1)*, 56-63.

University of Oxford (2006). Entwined in the network of networks (interview with Bill Dutton), in *Annual Review 2004/2005*, Oxford: Public Affairs Directorate, 6.

Vuori, S. & Holmlund-Rytkonen, M. (2005). 55+ people as internet users. *Marketing Intelligence & Planning, 23(1)*, 58-76.

Watt, D. & White, J. (1999). Computers and the family life. *Journal of Comparative Family Studies, 30*, 1-15.

In: Internet Issues: Blogging, the Digital Divide and Digital... ISBN: 978-1-60741-824-5
Editor: Cassie M. Evans, pp. 131-145

Chapter 8

DIGITAL OUTDISTANCING: SOCIO-ECONOMIC STATUS DIFFERENCES IN USE OF THE INTERNET, WHAT COULD IT MEAN FOR HEALTH DISPARITIES?

Silje C. Wangberg
Norwegian Centre for Telemedicine, University Hospital of North Norway

ABSTRACT

This chapter aims to explore relations between socio-economic status (SES), Internet use, social capital and subjective health. First, these relations are studied in the European Social Survey round 3 data set (ESS3; N = 33017). Participants were from representative samples between 15 and 80 years of age from 18 different European countries. In Study 1 it was found that SES is related to media use, in that those with a higher educational attainment and income had used the Internet more, and watched less TV. Furthermore, a simple correlation analysis showed that Internet use had positive relations to social capital variables and health, while TV has negative. When controlling for income, educational attainment, income and gender in hierarchical regression analysis, the positive relation between Internet use and subjective health remain significant but very small. The positive relation between Internet use and interpersonal trust and social interaction, respectively, were somewhat larger. A second study assessed the characteristics of users of an online smoking cessation intervention. Data on demographic characteristics of registered users in an Internet-based smoking cessation program were compared with data from an annual population smoking survey in conjunction with census data. In Study 2 it was found that the intervention had higher reach in strata of smokers that were younger, more highly educated and female. Finally, the inequalities within health online are discussed in light of the above.

INTRODUCTION

The concept 'digital divide' is used to describe a gap in physical access to the Internet. Consequently, policies are aimed at providing access for all. The European Council has for instance as part of their response to the economic downturn invested towards the goal of 100% coverage of broadband Internet between 2010 and 2013 (the Competitiveness Council, 2009). Currently, 60% of the EU households have Internet access, and for the northwest European countries, >80% of the households have access (Eurostat, 2008).

Closing the access gap, however, makes differences in the use of Internet even more visible. The current chapter deals with socio-economic (SES) and -demographic differences in use of the Internet for health purposes. Thus, I do not treat the digital divide as a dichotomous gap between the "haves" and the "have nots", but rather as a continuum in the opportunities and motivations to utilize the Internet for health purposes.

Studies of interventions attempting to close the digital divide illustrates that access is not enough. Ferlander and Timms (2007) studied the effects on social capital of two different interventions in a multi-cultural and socially disadvantaged city area in Sweden. One intervention provided free Internet access to everyone, but was hardly used, eventually closed down, and those who used it mainly used it for general high-speed access to the Internet rather than utilizing the local community portal. In contrast, the second intervention where Internet access was provided against a low cost in the context of a local café where informal support and face-to-face interaction opportunities also was available was successful in increasing social interactions both within and outside of the local area. Some studies also indicate that interventions to close the digital divide can lead to health benefits (Kreps, 2005; Shaw, et al., 2006).

Social Capital and Internet Use

It has long been known within health promotion that there is a great divide between those with access to health information and those without (Tones & Tilford, 2001). The core of the concept of 'social capital' rests in the availability of resources which can be accessed through social relationships (Fields, 2003). These social relationships may be both formal participation in organizations and more informal social interactions, which is the focus of this chapter. Trust can be seen as a prerequisite for participating in social networks, and at the same time a main outcome of such participation (Fields, 2003). Social capital is, according to Bourdieu (1986), an important means for perpetuating differences in social status. Along a similar vein, one might speculate that the Internet might be used to a greater extent to uphold and utilize social networks among those with a higher SES.

Social capital is linked to better health outcomes such as reduced mortality risk and higher self-rated health (Islam, Merlo, Kawachi, Lindström, & Gerdtham, 2006) as well as being negatively related to risk factors such as waist circumference and obesity (Moore, Daniel, Paquet, Dube, & Gauvin, 2009). Social capital can be divided into bonding ties which connect people within a social strata or local community, and bridging ties which connect people across social levels and distance (Fields, 2003). While bonding capital is related to health by among other things, providing emotional support, bridging capital is important in among other things, accessing health information (Ferlander, 2007). The Internet might provide opportunities for increasing both kinds of social capital (Hampton & Wellman, 2001, 2003; Shiau, 2008; Wellman, Haase, Witte, & Hampton, 2001).

Research has shown both negative and positive potentials of the Internet when it comes to social capital. On the one hand, there are time displacement effects, where people spending more time online necessarily spend less time on other activities, including face-to-face and phone interactions (Shklovski, Kraut, & Rainie, 2004; Wellman, et al., 2001). Nie and colleagues (e.g. Nie & Hillygus, 2002) have shown this nicely with time diary studies where they find that the more time spent online at home and during weekends, the less time spent face-to-face with family and friends.

On the other hand, the Internet may provide an excellent tool to maintain and extend social networks in less time and regardless of location (Wellman, et al., 2001). Many relationships formed online will be extended to "real life" and may also hold higher levels of intimacy faster than those formed off-line (Bargh & McKenna, 2004). The Internet also facilitates the search for others sharing specific interests, creating bonding social capital. However, although certain online groups host discussions of particular interest, they also gather diverse people, increasing bridging social capital on many levels, i.e. social capital, which connects actors to resources, relationships and information beyond their immediate environment (Ferlander & Timms, 2007). People can thus use the Internet both to maintain and strengthen existing social capital both of the bonding and bridging kind, as well as gaining new ties both within and across groups. From an individual perspective, the social network(s) online might thus constitute capital for health through access to both professional health information as well as sharing peer experiences on health services, treatments and symptoms.

Socio-Economic Status (SES) and Health-Related Internet Use

Earlier studies both in Norway (Andreassen, Wangberg, Wynn, Sørensen, & Hjortdahl, 2006) and elsewhere (Andreassen, et al., 2007; Fox, 2006; Lorence, Park, & Fox, 2006; Ybarra & Suman, 2006) have shown that women use the Internet for health purposes to a greater extent than men. There are, however, previous studies that do not find this gender difference (Andreassen, Sandaune, Gammon, & Hjortdahl, 2002; Närhi & Helakorpi, 2007). A more consistent finding is that use of the Internet for health purposes is higher amongst those with higher education (Andreassen, et al., 2007; Fox, 2006; Lorence, et al., 2006; Ybarra & Suman, 2006) and amongst younger people (Andreassen, et al., 2007). Surveys with patient populations also confirm these demographic trends in use of the Internet for health purposes (Rahmqvist & Bara, 2007; Schwartz, et al., 2006).

A European eHealth Consumer Trends Survey was carried out in seven European countries in October-November 2005 and repeated in April-May 2007. In the first round (Andreassen, et al., 2007), it was found that use of the Internet for health purposes was higher amongst those with higher education, women and amongst younger people. In the second round (Kummervold, et al., 2008), a significant growth in the use of, as well as the rated importance of, the Internet for health purposes was found in all the seven countries. Although gender differences in general use of the Interned had diminished, young women were still being the most active users of the Internet for health purposes. Furthermore, students or employed used the Internet significantly more both in general and for health purposes than those unemployed.

Wangberg et al (2008) used the same dataset as Andressen et al (2007) to compare Internet users who had used the Internet for health purposes with Internet users who had not used it for health purposes. They found that although most differences were small, those who had used the Internet but not for health purposes tended towards lower educational attainment, more children, visiting their GP less often, and were more likely to be male. Thus, there are digital divides in how the Internet is used, potentially contributing to further inequalities in health.

Using data from the second round of the European Social Survey (ESS2), Wangberg et al (2008) found that Internet use was found to be more closely related to social support and subjective health than use of other media. Internet use was also found to be a plausible mediator between SES and subjective health, especially through interacting with social support.

The first study reported in the current chapter replicates and extends these findings through using data from the third round of the European Social Survey (ESS3) to assess the relations between SES, Internet use, social capital and subjective health. The second study investigates characteristics of the sample of smokers reach with an Internet-based smoking cessation intervention in Norway.

Study 1. Relations between SES, Internet Use, Social Capital and Subjective Health

Methods Study 1

The sample consisted of people from Austria, Belgium, Bulgaria, Cyprus, Germany, Denmark, Estonia, Spain, Finland, France, United Kingdom, Ireland, Netherlands, Norway, Poland, Sweden, Slovenia and Slovakia (N= 33017; for more details see www. europeansocialsurvey.org). Latvia and Romania were excluded because of incomplete design weights, while Ukraine, Switzerland and Hungary were completely lacking income information, and Portugal had more than 50% missing data on household income. The Russian Federation was excluded as they had a much larger sample size than the other countries and 80% of this sample did not have access to or did not use the Internet.

Data were gathered on the following variables: gender, age, education as measured through the International Standard Classification of Education (ISCED; UNESCO, 1997), total household income, Internet use, subjective health status, interpersonal trust and social interactions.

Internet use was measured with the item: “how often do you use the internet, the World Wide Web or e-mail – whether at home or at work –for your personal use?” Response alternatives were: “No access at home or work”, “Never use”, “Less than once a month”, “Once a month”, “Several times a month”, “Once a week”, “Several times a week”, or “Every day”. Internet use was thus measured and treated in analyses in accordance with the notion of a digital continuum rather than a dichotomous divide. Use of TV and Newspapers was rated on an 8-point scale ranging from “No time at all” to “more than 3 hours” on “an average weekday". Subjective health was measured with one item, ''How is your health in general? Would you say it is 5 = very good, 4 = good, 3 = fair, 2 = bad, or 1 = very bad?''. The scale was reversed before analysis.

Interpersonal trust was measured with three statements: other people can be trusted, people are generally fair, and people are generally helpful. Contrary to other surveys, a linear, numeric scale of 1 to 10 is used. Social interaction was measured with three items: 1) "how often do you meet socially with friends, relatives or work colleagues?". The response alternatives were the same as for Internet use. 2) "Compared to other people of your age, how often would you say you take part in social activities?". The response alternatives were: "Much less than most", "Less than most", "About the same", "More than most", or "Much more than most". The 3rd question "do you have anyone to discuss intimate issues with" was weighted so that yes=7 and no=0 to harmonize with the scale of the other two social interaction questions.

The statistical software package SPSS was used to perform correlations between all variables and hierarchical regression analyses with subjective health and the two social capital variables, respectively, as dependent variables and Internet use and socio-economic and – demographic variables as predictors. The expectation-maximum likelihood algorithm was used to impute missing data (Weichung, 1989). Two weights were combined and applied for the analyses: a design weight adjusting for deviations in sampling in some of the countries, and a population weight adjusting for population size differences between the countries, and thus reflecting the fact that although some countries in Europe, like the Scandinavian, have high Internet use rates, the absolute number of people using the Internet combined in Europe is lower (for more details on weighting see ESS).

Results Study 1

Overall, 44% of the respondents reported not having access to or never having used the Internet, ranging from 77% in Bulgaria to 20% in Norway. Conversely, 50% of Danes reported daily Internet use compared to 10% of Cypriots. As can be seen in Table 1, the distributions are split, with an overweight of people either not using the Internet at all, or using it daily.

Table 2 shows correlations between the measured variables. Here we can see that socio-economic variables are related to media use. Internet use is most strongly positively related to income, and secondly to education. Older people and females tend to use the Internet less. For TV watching the correlations are weaker, and in the opposite direction. Higher education is related to less TV watching, and older people watch more TV.

For health and the social capital variables we see that Internet use is positively related to subjective health, social interactions, and interpersonal trust. TV watching on the other hand is weakly and negatively related to all of these variables.

But when controlling for income, educational attainment, income and gender in hierarchical regression analysis, we see in Table 3 that the positive relation between Internet use and subjective health remain significant but very small. The two social capital variables had a weak relationship, but contributed significantly to the full model that accounted for 22% of variance in subjective health.

Table 1. Personal use of the Internet in 18 European countries (weighted N=30660)

	Percent
No access at home or work	30.7
Never use	13.5
Less than once a month	2.2
Once a month	2.0
Several times a month	4.3
Once a week	4.7
Several times a week	14.0
Every day	28.6

Table 2. Correlations with between media use, subjective health, social capital and socio- economic status in the ESS3 (weighted N = 30660)

	Radio listening	Newspaper reading	Internet use	Health	Female	Age	Edu	Income	Trust	Social interact
TV watching	.020**	.089**	-.231**	-.137**	.021**	.212**	-.191**	-.151**	-.045**	-.076**
Radio listening		.095**	.026**	.037**	.002	-.009	.018**	.045**	.027**	.027**
Newspaper reading			.005	.012*	-.083**	.199**	.128**	.090**	.088**	.013*
Internet use				.297**	-.101**	-.462**	.419**	.452**	.157**	.235**
Health					-.056**	-.372**	.207**	.291**	.187**	.212**
Female						.036**	-.033**	-.086**	.020**	.010
Age							-.157**	-.161**	.016**	-.229**
Education								.388**	.164**	.109**
Income									.272**	.177**
Trust										.166**

**. Correlation is significant at the 0.01 level (2-tailed).

*. Correlation is significant at the 0.05 level (2-tailed).

Internet use had a somewhat larger relation to interpersonal trust and social interaction, respectively, when controlling for SES variables.

Discussion Study 1

In line with the findings in Study 1, several studies have found a positive relation between Internet use and interpersonal trust (Beaudoin, 2008; Gardner & Oswald, 2001; Stark, 2009). Although I did not test for such an interaction effect, another European study (Zinnbauer, 2007) have found that the Internet mostly facilitates building of social capital among those with a higher SES.

The study has some limitations. A limitation that Study 1 shares with many others is it's cross-sectional nature. For instance, it is just as likely that people who are highly social use the Internet more as it is that people become more social from using the Internet. A longitudinal study on these issues found few and weak effects of Internet use on other forms of social interaction (Shklovski, Kraut, & Rainie, 2004a).

Table 3. Hierarchical regression on ESS3 data with subjective health as dependent variable and demographic variables and Internet use, as predictors (weighted N = 30637)

		B	95% CI		β	t	P	R^2
1	Female	-.062	-.026	-.044	-.024	-4.72	<.001	.199
	Age	-.017	-.016	-.016	-.326	-62.63	<.001	
	Education	.039	.052	.046	.075	13.44	<.001	
	Income	.078	.086	.082	.208	37.07	<.001	
2	Female	-.059	-.022	-.041	-.022	-4.37	<.001	.199
	Age	-.016	-.015	-.016	-.314	-54.38	<.001	
	Education	.034	.048	.041	.067	11.54	<.001	
	Income	.074	.083	.078	.198	33.41	<.001	
	Internet use	.006	.014	.010	.032	4.86	<.001	
3	Female	-.047	-.065	-.029	-.026	-5.10	<.001	.208
	Age	-.015	-.015	-.014	-.299	-51.44	<.001	
	Education	.041	.034	.048	.068	11.70	<.001	
	Income	.074	.070	.079	.188	31.76	<.001	
	Internet use	.006	.002	.010	.020	3.07	.002	
	Social Interaction	.029	.026	.032	.098	18.50	<.001	
4	Female	-.056	-.074	-.038	-.031	-6.14	<.001	.222
	Age	-.016	-.016	-.015	-.313	-54.09	<.001	
	Education	.037	.030	.043	.060	10.53	<.001	
	Income	.064	.059	.068	.161	26.85	<.001	
	Internet use	.004	.000	.008	.013	1.96	.049	
	Social Interaction	.024	.021	.027	.124	15.39	<.001	
	Trust	.021	.019	.023	.082	23.25	<.001	

Table 4. Hierarchical regression on ESS3 data with social interaction as dependent variable and demographic variables and Internet use, as predictors (weighted N = 30637)

		B	95% (CI)		β	t	P	R^2
1	Female	.118	.249	.183	.030	5.48	<.001	.074
	Age	-.036	-.032	-.034	-.204	-36.34	<.001	
	Education	.025	.073	.049	.024	4.03	<.001	
	Income	.167	.198	.182	.138	22.88	<.001	
2	Female	.158	.289	.223	.037	6.70	<.001	.083
	Age	-.028	-.024	-.026	-.158	-25.52	<.001	
	Education	-.036	.013	-.012	-.006	-0.92	.358	
	Income	.119	.152	.135	.102	16.07	<.001	
	Internet use	.108	.136	.122	.122	17.09	<.001	

This study was to a large extent limited to individual level analysis, but both social capital and Internet use should of course also be seen in light of macro variables. Räsänen showed a relatively large effect of country of the respondent on Internet use, even after controlling for income and education, with the Nordic countries being most strongly positively related to Internet use (Räsänen, 2006). Social capital is also higher in the Nordic welfare states (Fields, 2003; Islam, et al., 2006). Islam et al found in their review that the relation between health and social capital in these Nordic countries did not correlate highly within neighbourhoods, suggesting that the local communities are not the main source of the high social capital in these countries. Huang et al (2003) found a high correlation at the level countries between interpersonal trust and Internet penetration.

Another important covariate that was not controlled for in the studies reported in this chapter was broadband Internet access, as those with high-speed access utilize the Internet more for health purposes (Rains, 2008). Furthermore, broadband access could interact with socio-economic factors. For instance, in Norway, which is a country with high broadband coverage as well as relatively small social differences, 73% of all households have high-speed access, but the percentage that has broadband connections increases with increasing income (Statistics Norway, 2008).

Study 2. Who Quits Online? Comparison of User Demographics for an Internet-Based Smoking Cessation Program with the Norwegian Population of Smokers

This study presents characteristics of the users that signed up for the Internet-based smoking cessation program www.slutta.no during the first six months after the site was opened, and compares this sample with the total populations of Norwegian smokers.

Methods Study 2

Data on the users of the Internet-based smoking cessation program www.slutta.no was gathered from August 2006 until February 2007 as part of a randomized controlled trial. Demographic characteristics of all the 1902 users that registered in this period are reported in this study and compared data on the population of Norwegian smokers between the age of 16 and 74. Informed consent was attained from all users at the beginning of the registry process.

The main outcome reported is estimates of proportions of daily smoker population strata reached. These estimates were based an annual population smoking survey in conjunction with census data.

Data on the population of Norwegian smokers between the age of 16 and 74 was gathered by Statistics Norway (2) on behalf of the Directorate for Health and Social Affairs. Over the year 2006, four surveys were carried out, resulting in a final sample of 4756. The main survey questions asked were "Do you sometimes smoke?" and "Do you smoke daily or occasionally?" In addition, information was gathered on age, gender and educational attainment. This study uses data on demographic characteristics of those reporting that they smoked daily.

Results Study 2

A total of 1903 participants had registered for the Internet-based smoking cessation program Opptur between August 2006 and February 2007. Data on age or gender was missing for 60 participants (3.2 %). In addition, 12 participants were younger than 16, ten of them females of 15 years of age, leaving a valid sample of 1842 for the population comparisons.

The total estimated reach of the intervention during the first six months was 0.2 % of the Norwegian daily smokers.

Overall a larger proportion of the population of female daily smokers were recruited to the program during the first six months than of the male population. Likewise, higher proportions of daily smokers younger than 44 years where recruited than amongst those older. The highest reach of smokers was amongst women aged 25-34 where 0.63 % of the population of daily smokers were recruited to the program during the first 6 months. The lowest reach was in the age group 65-74 years where for both men and women 0.01 % of the respective populations of daily smokers were recruited. All proportions can be seen in Figure 1.

The intervention reached 0.96 % of the population of daily smokers with higher education, compared with only 0.10 of those with secondary schooling and 0.04 % of daily smokers that had only completed primary education. The highest reach was amongst daily smokers with a higher education aged 16-25 where 5.27 % of the population of daily smokers were recruited to the program during the first 6 months. The lowest reach was amongst daily smokers that had completed only primary education in age group 65-74 years where none were recruited. All proportions can be seen in Figure 2.

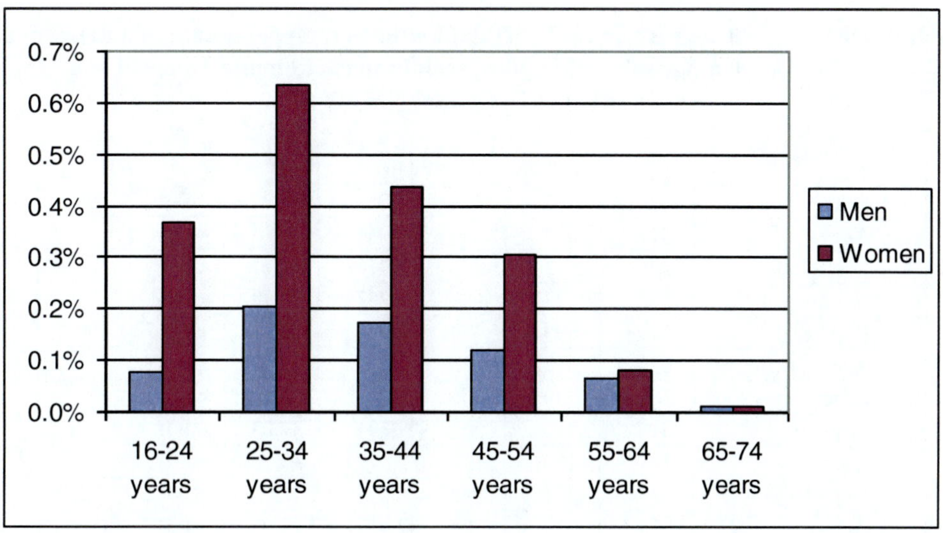

Figure 1. Proportions of daily smokers in Norway reached during the first six months after launching an Internet-based smoking cessation program by gender and age

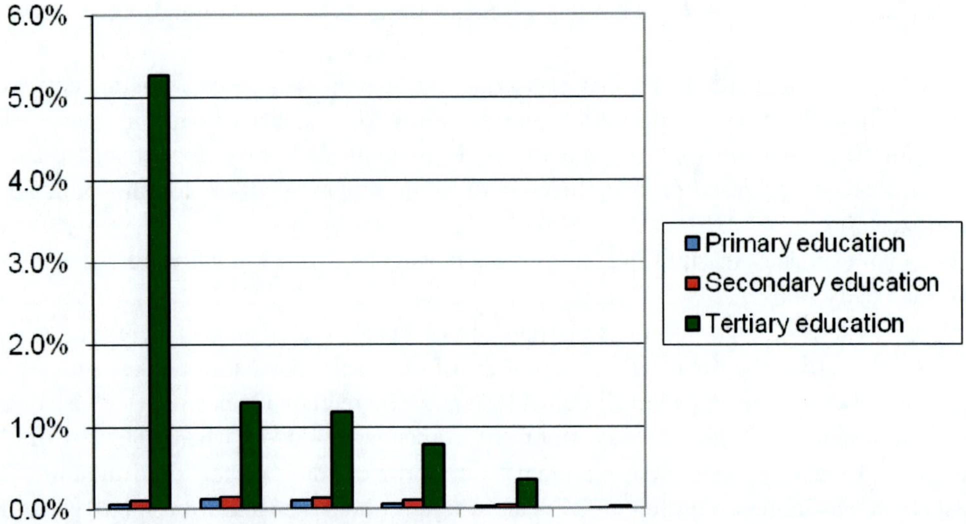

Figure 2. Proportions of daily smokers in Norway reached during the first six months after launching an Internet-based smoking cessation program by educational attainment and age

Discussion Study 2

The smoking cessation intervention had the greatest reach amongst younger smokers, females and those with higher education. Thus replicating the general pattern of the eHealth consumer presented earlier in this chapter.

Table 5. Hierarchical regression on ESS3 data with as interpersonal trust dependent variable and demographic variables, social interaction and Internet use, as predictors (weighted N = 30637)

		B	95% CI		β	t	P	R^2
1	Female	.335	.565	.450	.042	7.67	<.001	.290
	Age	.017	.023	.020	.068	12.13	<.001	
	Education	.231	.315	.273	.076	12.77	<.001	
	Income	.573	.627	.600	.257	42.88	<.001	
2	Female	.379	.609	.494	.046	8.42	<.001	.296
	Age	.025	.032	.028	.096	15.61	<.001	
	Education	.162	.249	.206	.057	9.25	<.001	
	Income	.519	.577	.548	.234	37.02	<.001	
	Internet use	.111	.160	.135	.077	10.76	<.001	
3	Female	.328	.556	.442	.041	7.59	<.001	.322
	Age	.031	.038	.034	.117	18.99	<.001	
	Education	.165	.252	.209	.058	9.45	<.001	
	Income	.487	.545	.516	.221	35.04	<.001	
	Internet use	.082	.131	.107	.060	8.52	<.001	
	Social interaction	.215	.254	.235	.133	23.58	<.001	

Other studies of online smoking cessation forums have found that women post the majority of messages, and that these messages from females tend to be more emotionally supportive (Burri, Baujard, & Etter, 2006). Reciprocity is regarded as a crucial part of creating social capital and trust. Reciprocity is probably even more crucial for online communities specifically.

One of the reasons that men and people with higher education use the Internet more probably has to do with higher Internet skills (Hargittai, 2002; Hargittai & Shafer, 2006) and online information processing skills more specifically (van Deursen & van Dijk, 2009). Hargittai and Shafer (2006) found that while education was related to actual Internet skill, it was not related to perceived skills. Gender on the other hand was related to perceived skills but not actual skills, with females perceiving themselves as less skilled than they actually were (Hargittai & Shafer, 2006). This could explain some of the male dominance in general Internet usage, although it does not explain the dominance of females in health-related Internet use.

Birru and colleagues (2004) showed some of the problems encountered by low-literacy adults when trying to find health information online. The participants had problems specifying effective search terms, tended to prefer sponsored links (which lead to alternative cancer treatments), and ended up in websites with a high readability level. Many of the participants could identify and read back the relevant information on these sites, but were unable to paraphrase the information in their own words, suggesting limited comprehension of the material. Furthermore, this study pointed out some of the challenges in involving the target population in studies to improve eHealth tools, as they had trouble understanding the informed consent form, as well as participating in a think-a-loud session.

While Zillien and Hargittai (2009) found a clear SES gradient in using the Internet to read political news or stock quotes, the effect was less pronounced for health, but still significant. Of those with the highest SES, 45% reported searching for health information, compared to 29% in the lowest SES group, and 40% across all seven status groups. The status effect on health related Internet use did, however, become insignificant when controlling for age, gender and interest in topic, all which were significant predictors of health related Internet use (Zillien & Hargittai, 2009). That interest in health as a moderator between SES and health related Internet use is not so surprising from a functionalistic account of media use; people with higher SES are to a greater extent able to utilize the Internet to meet their needs, whether it is politics, stocks or health.

As for health specific usage, it seems likely that both an interest in health topics and a sufficient level of health literacy is needed. Dutta-Bergman (2005) did not, however, find a relation between health consciousness and Internet use. He did, however, find a small positive relation between seeking additional health information beyond the primary physician and Internet use (Dutta-Bergman, 2005). Rains (2007), on the other hand, found that attention to health information predicted health-related Internet use. Furthermore, he found that both trust in the Internet as a source of cancer information and trust in the doctor were positively related to searching the Internet for cancer information.

Health Literacy is the skills needed by the individual to gain access to, understand and use information in ways that promote and maintain good health (Nutbeam, 1998). Thus, health literacy means more than being able to read and understand information on a website. Fundamental cognitive functioning and general literacy is necessary, but not sufficient. Nutbeam (Nutbeam, 2000) has divided health literacy into three types: 1) Basic/functional literacy, which denotes sufficient basic skills in reading and writing to be able to function effectively in everyday situations; 2) Communicative/interactive literacy, which denotes more advanced cognitive and literacy skills which, together with social skills, can be used to actively participate in, extract information from and derive meaning from different forms of communication, and to apply new information to changing circumstances; and 3) Critical literacy, which denotes cognitive and social skill to critically analyse information, and to use this information to exert greater control over life events and situations, such as engaging in shared decision making. Some research groups have made even finer distinctions than this, but sufficient to say that the skills needed for successfully filling the roles of the active or empowered patient online are diverse, including various self-efficacies. For intervention development this means that in addition to ensuring that text is tailored to reading ability and/or education level, other skills need to be assessed and, if needed, improved. Delivering an intervention online poses additional demand on the user, both in terms of confidence and skills. This might be called eHealth literacy (Norman & Skinner, 2006).

CONCLUSION

The smoking cessation intervention showed that reach was skewed in terms of socio-demographic characteristics, following the same patterns as the survey studies reviewed earlier in this chapter. Therefore it seems reasonable to conclude that health behavioural interventions offered online may well run the risk of empowering the empowered.

We saw that there is a visible divide between people not using the Internet at all and those using it daily, and not only between those that have access or not. This divide, however, did seem to matter more directly for social capital variables rather than health. In some ways health in Western societies has become more of an individual matter, rather than a collective one. Individuals in this sense become more dependent on networks beyond the family or local group in order to gain access to relevant resources such as information about health behaviours or how to navigate within organizations to obtain care. The SES effects on social capital and interaction might in this sense interact towards greater health inequalities.

Future research efforts should be aimed at finding ways to minimize the risk that Internet-based health interventions contribute to increasing social disparities in health.

REFERENCES

Andreassen, H. K., Bujnowska-Fedak, M. M., Chronaki, C. E., Dumitru, R. C., Pudule, I., Santana, S., et al. (2007). European citizens' use of E-health services: A study of seven countries. *BMC Public Health, 7(53)*.

Andreassen, H. K., Sandaune, A. G., Gammon, D. & Hjortdahl, P. D. (2002). Use of Internet health services in Norway. *Journal of the Norwegian Medical Association, 122(17)*, 1640-1644.

Andreassen, H. K., Wangberg, S. C., Wynn, R., Sørensen, T. & Hjortdahl, P. (2006). Health-related use of the Internet in the Norwegian population. *Journal of the Norwegian Medical Association, 126*, 2950-2952.

Bargh, J. & McKenna, K. (2004). The Internet and social life. *Annual Review of Psychology, 55*, 573-590.

Beaudoin, C. E. (2008). Explaining the Relationship between Internet Use and Interpersonal Trust: Taking into Account Motivation and Information Overload. *Journal of Computer-Mediated Communication, 13(3)*, 550-568.

Birru, S. M., Monaco, M. V., Charles, L., Drew, H., Njie, V., Bierria, T., et al. (2004). Internet Usage by Low-Literacy Adults Seeking Health Information: An Observational Analysis. *J Med Internet Res, 6(3)*, e25.

Bourdieu, P. (1986). The forms of capital. . In: J. Richardson, (Eds.), *Handbook of Theory and Research for the Sociology of Education* (241-258). New York: Greenwood Press.

Burri, M., Baujard, V. & Etter, J. F. (2006). A qualitative analysis of an Internet discussion forum for recent ex-smokers. *Nicotine & Tobacco Research, 8*(1 supp), 13-19.

Dutta-Bergman, M. (2005). Developing a profile of consumer intention to seek out additional information beyond a doctor: The role of communicative and motivation variables. *Health Communication, 17(1)*, 1-16.

ESS Weighting European Social Survey Data, *Design Documentation* (1-5).

Eurostat (2008). Level of Internet access - households - Percentage of households who have Internet access at home. Retrieved July 10, 2009: http://epp.eurostat.ec.europa.eu/tgm/table.do?tab=table&init=1&plugin=1&language=en&pcode=tsiir040.

Ferlander, S. (2007). The importance of different forms of social capital for health. *Acta Sociologica, 50(2)*, 115.

Ferlander, S. & Timms, D. (2007). Social Capital and Community Building through the Internet: A Swedish Case Study in a Disadvantaged Suburban Area. *Sociological Research Online, 12(5)*.

Fields, J. (2003). *Social Capital (Key Ideas)*: Routledge New York.

Fox, S. (2006). *Online Health Search 2006: Most internet users start at a search engine when looking for health information online. Very few check the source and date of the information they find.*: Pew Internet & American Life Project.

Gardner, J. & Oswald, A. (2001). Internet use: the digital divide. In: A., Park, J., Curtice, K., Thomson, L., Jarvis, C. Bottomley, & N. Stratford, (Eds.), *British Social Attitudes: the 18th Report: Public Policy, Social Ties* (159-173). London: Sage.

Hampton, K. & Wellman, B. (2001). Long Distance Community in the Network Society: Contact and Support Beyond Netville. *American Behavioral Scientist, 45(3)*, 476-495.

Hampton, K. & Wellman, B. (2003). Neighboring in Netville: How the Internet Supports Community and Social Capital in a Wired Suburb. *City & Community, 2(4)*.

Hargittai, E. (2002). Second-level digital divide: Differences in people's online skills. *First Monday, 7(4)*.

Hargittai, E. & Shafer, S. (2006). Differences in Actual and Perceived Online Skills: The Role of Gender. *Social Science Quarterly, 87(2)*, 432-448.

Huang, H., Keser, C., Leland, J. & Shachat, J. (2003). Trust, the Internet, and the digital divide. *IBM Systems Journal, 42(3)*, 507-518.

Islam, M., Merlo, J., Kawachi, I., Lindström, M. & Gerdtham, U. (2006). Social capital and health: Does egalitarianism matter? A literature review. *International Journal for Equity in Health, 5(1)*, 3.

Kreps, G. L. (2005). Disseminating relevant health information to underserved audiences: implications of the Digital Divide Pilot Projects. *J Med Libr Assoc, 93*(4 supl).

Kummervold, P. E., Chronaki, C. E., Lausen, B., Prokosch, H. U., Rasmussen, J., Santana, S., et al. (2008). eHealth trends in Europe 2005-2007: a population-based survey. *J Med Internet Res, 10(4)*, e42.

Lorence, D. P., Park, H. & Fox, S. (2006). Assessing health consumerism on the Web: A demographic profile of information-seeking behaviors. *Journal of Medical Systems, 30*, 251-258.

Moore, S., Daniel, M., Paquet, C., Dube, L. & Gauvin, L. (2009). Association of individual network social capital with abdominal adiposity, overweight and obesity. *Journal of Public Health.*

Nie, N. & Hillygus, D. (2002). The impact of Internet use on sociability: Time-diary findings. *IT & Society, 1(1)*, 1-20.

Norman, D. C. & Skinner, A. H. (2006). eHealth Literacy: Essential Skills for Consumer Health in a Networked World. *J Med Internet Res, 8(2)*, e9.

Nutbeam, D. (1998). Health promotion glossary. *Health Promot. Int., 13(4)*, 349-364.

Nutbeam, D. (2000). Health literacy as a public health goal: a challenge for contemporary health education and communication strategies into the 21st century. *Health Promot. Int., 15(3)*, 259-267.

Närhi, U. & Helakorpi, S. (2007). Sources of medicine information in Finland. *Health Policy, doi:10.1016/j.healthpol. 05.005.*

Rahmqvist, M. & Bara, A. C. (2007). Patients retrieving additional information via the Internet: A trend analysis in a Swedish population, 2000-05. *Scandinavian Journal of Public Health, 35(5)*, 533-539.

Rains, S. (2007). Perceptions of traditional information sources and use of the World Wide Web to seek health information: Findings from the health information national trends survey. *Journal of Health Communication, 12(7)*, 667-680.

Rains, S. (2008). Health at High Speed. *Communication Research, 35(3)*, 283-297.

Räsänen, P. (2006). Information society for all? Structural characteristics of Internet use in 15 European countries. *European Societies, 8(1)*, 59-81.

Schwartz, K. L., Roe, T., Northrup, J., Meza, J., Seifeldin, R. & Neale, A. V. (2006). Family Medicine Patients' Use of the Internet for Health Information: A MetroNet Study. *J Am Board Fam Med, 19(1)*, 39-45.

Shaw, B., Gustafson, D. H., Hawkins, R., McTavish, F., McDowell, H., Pingree, S., et al. (2006). How Underserved Breast Cancer Patients Use and Benefit From eHealth Programs: Implications for Closing the Digital Divide. *American Behavioral Scientist, 49(6)*, 823-834.

Shiau, H. (2008). Revisiting'bridging or binding': social capital and the uses of Instant Messenger (IM) and blog. *International Journal of Electronic Democracy, 1(1)*, 85-97.

Shklovski, I., Kraut, R. & Rainie, L. (2004). The Internet and social participation: Contrasting cross-sectional and longitudinal analyses. *Journal of Computer-Mediated Communication, 10(1)*, 1-23.

Stark, A. (2009). *A study of Internet use and its impact on individual Level social capital indicators and motivation to volunteer.* University of Missouri.

Statistics Norway (2008). *The Information Society.* Retrieved from http://www.ssb.no/english/subjects/10/03/ikt_en/.

Key Issues Paper 2009 - contribution from the Competitiveness Council to the Spring European Council.

Tones, K. & Tilford, S. (2001). *Health Promotion: Effectiveness, Efficiency and Equity* (3rd ed.). Cheltenham: Nelson Thornes.

UNESCO (1997). *International Standard Classification of Education*. Retrieved from http://www.unesco.org/education/information/nfsunesco/doc/isced_1997.htm.

van Deursen, A. & van Dijk, J. (2009). Improving digital skills for the use of online public information and services. *Government Information Quarterly, 26(2)*, 333-340.

Wangberg, S. C., Andreassen, H. K., Prokosch, H. U., Santana, S. M., Sorensen, T. & Chronaki, C. E. (2008). Relations between Internet use, socio-economic status (SES), social support and subjective health. *Health Promot Int, 23(1)*, 70-77.

Weichung, J. (1989). Influence of incomplete observations in multiple linear regression. *Statistics & Probability Letters, 8*, 171-174.

Wellman, B., Haase, A. Q., Witte, J. & Hampton, K. (2001). Does the Internet Increase, Decrease, or Supplement Social Capital?: Social Networks, Participation, and Community Commitment. *American Behavioral Scientist, 45(3)*, 436-455.

Ybarra, M. L. & Suman, M. (2006). Help seeking behavior and the Internet: A national survey. *International Journal of Medical Informatics, 75(1)*, 29-41.

Zillien, N. & Hargittai, E. (2009). Digital Distinction: Status-Specific Types of Internet Usage. *Social Science Quarterly, 90(2)*, 274-291.

Zinnbauer, D. (2007). *What can Social Capital and ICT do for Inclusion*. Institute for Prospective Technological Studies. Seville: European Commission Joint Research Centre.

In: Internet Issues: Blogging, the Digital Divide and Digital… ISBN: 978-1-60741-824-5
Editor: Cassie M. Evans, pp. 147-164

Chapter 9

SHUT UP AND LISTEN: THE INFLUENCE OF SELECTIVE EXPOSURE TO BLOGS ON POLITICAL TOLERANCE

***Thomas J. Johnson**, *Shannon L. Bichard*‡ *and Weiwu Zhang*≠**
Texas Tech University, Lubbock, TX, USA

ABSTRACT

The Internet has provided scholars a unique opportunity to address the issue of democracy and political polarization by exploring the relationship between political tolerance and selective exposure. This study uses an online panel of Internet users from a major university to explore the degree to which selective exposure to blogs predicts tolerance to viewpoints they oppose. Specifically, this study investigates the degree to which selective exposure as well as reliance on blogs influences political tolerance after controlling for political and demographic factors. Reliance on face-to-face discussion is also examined as a source of direct and indirect influence. The results indicate that both selective exposure and reliance on blogs have no effect on levels of political tolerance; while increased political activity, higher socioeconomic status, and reliance on face-to-face discussion appear to exert a positive influence on political tolerance. Future research should continue examining specific factors that predict political tolerance in order to better understand and preserve a democratic climate conducive to freedom of expression.

INTRODUCTION

Communication scholars have increasingly worried about the tremendous growth in partisan content and citizen's increased use of such content in making political decisions.

* Marshall and Sharleen Formby Regents Professor, College of Mass Communication, Texas Tech University, Lubbock, TX 76409, t.johnson@ttu.edu, 806-742-6500 ext. 253
‡ Associate Professor, Shannon.Bichard@ttu.edu
≠ Assistant Professor, Weiwu.zhang@ttu.edu

Some researchers contend that partisan selective exposure, defined as seeking out information that supports one's point of view and avoiding information that challenges it, will lead to more polarized attitudes and more fragmented political views (Knoblach-Westerwick & Meng, 2009; Iyengar & Hahn, 2009; Stroud, 2006; Sunstein, 2001, 2007). Polarized views could lead to people processing information in a more biased manner (Meffert, Chung, Joner, Walks & Garst, 2006) and create less tolerance of opposing viewpoints (Mutz, 2002). Selective exposure may also produce a less informed electorate as democratic theory is based on the premise that before citizens reach decisions they should gather and critically evaluate a range of opinions (Delli Carpini & Keeter, 1996; Stroud, 2006).

Political tolerance, then, is a lynchpin to the argument that selective exposure endangers the political process. Those who practice selective exposure will become less tolerant of opposing viewpoints, leading to increased polarization in society. Higher levels of selective exposure are linked to less tolerance of opposing viewpoints, which leads to increased polarization. However, few scholars have directly explored whether selective exposure to partisan content leads people to be less tolerant, and those studies have produced mixed results. For instance while many have found that those who are less tolerant of opposing points of view are more likely to practice selective exposure (Hardy, Scheufele & Wang, 2005; Mutz, 2002; Pettigrew, 1997), a recent study of the influence of political tolerance on selective exposure to blogs and political Websites found the opposite: Those who were more tolerant of opposing viewpoints were more likely to practice selective exposure (Johnson, Zhang & Bichard, 2009).

While scholars have increasingly examined the degree to which people practice selective exposure on the Internet (Bimber & Davis, 2003; Garrett, 2006; Graf & Aday, 2008; Mutz & Martin, 2001: Stroud, 2006, 2007, 2008), fewer studies have explored the degree to which people practice selective exposure to content on blogs, which are high in partisan content (Johnson, Kaye, Bichard & Wong, 2007; Johnson & Kaye, 2004, 2006; Johnson et al., 2009; Wall, 2006).

This study employs an online panel of Internet users from a major university to explore the degree to which selective exposure to blogs predicts tolerance to groups whose ideology they oppose. More specifically, this study examines the degree to which selective exposure to blogs directly predicts political tolerance as well as indirectly predicts tolerance through respondents' relationship with interpersonal political discussion after controlling for political and demographic factors.

POLITICAL TOLERANCE

Political tolerance is considered one of the psychological underpinnings of democracy. For democracy to flourish, a consensus must exist among the public about the rules of the game. One of those rules is that people must be able to freely express their views, and for such an environment to exist people must be tolerant of a range of political views, including ones they are personally opposed to (Sullivan & Transue, 1999; Sullivan, Piereson, & Marcus, 1982).[1] Indeed, classic liberal theory stresses individual rights, tolerance of

[1]Researchers have noted that in speaking of tolerance, one needs to distinguish between political tolerance and social tolerance. Political tolerance refers to the willingness to extend rights to, and support political actions

differences, and the preservation of civil rights as more important than public safety and national security (Keum, et al., 2005). But studies suggest that while there is fairly wide public support for abstract principles of tolerance, Americans are increasingly intolerant about accepting opposing views in practice (Gibson, 1992; Sullivan et al, 1982).

Scholars first became interested in studying political tolerance during the McCarthy Era when those who held views considered contrary to democratic values came under attack from the media and political elites (Sullivan & Transue, 1999). Not surprisingly, two extensive national surveys in the mid-1950s revealed that an overwhelming majority of Americans did not support civil liberties for socialists, atheists and communists (Stouffer, 1955). Similarly, Protho and Grigg (1960) discovered that while a majority of individuals approved the principles of democracy, the consensus disappeared when they were applied to specific controversial situations. These studies raised consternation among democratic theorists who wondered how democracy could survive when its basic tenets were not supported by the public.

Studies during the 1960s and early 1970s showed a marked improvement among support for political tolerance as the civil rights movement led to increasing support for minority rights and fear of Communism declined (Mueller, 1988). However, research has found that political tolerance decreased precipitously in wake of the events of 9/11 when studies indicated that two-thirds of Americans would be willing to sacrifice some civil liberties to combat terrorism and at least a quarter of Americans did not feel that the Bush Administration went far enough to restrict civil liberties in the months following 9/11 (Etzioni, 2002; Huddy, Khatid, & Capelos, 2002).

Studies have identified at least four sets of factors that influence the degree to whether individuals will be politically tolerant: background characteristics (including education, age, income, religiosity, partisanship, efficacy and trust) political expertise (which arises from political interest, activity and knowledge), internalization of democratic values and perceptions of threat (Gibson, 1986; Sullivan, et al., 1982: Sullivan & Transue, 1999). Media use has been linked to each of these four factors and this study focuses on how a particular type of media use, selective exposure, influences political tolerance

POLITICAL TOLERANCE AND SELECTIVE EXPOSURE

While some scholars cast doubt on whether the general population has become more polarized (Fiorina, Abrams, & Pope, 2005; Fiorina & Abrams, 2008), others have noticed more divisions within the country (Baldassarri, & Bearman, 2007; Abramowitz & Saunders, 2005) as more people position themselves on the end of the liberal/conservative continuum

of, groups that one finds abhorrent. On the other hand, social tolerance examines the willingness to socially engage people from groups they dislike (Markus, 2001). Researchers have also distinguished between support for freedom of expression and political tolerance, both of which are subsets of a support for general civil liberties. While political tolerance and support for freedom of expression are moderately correlated, they are not identical. Tolerance indicates acceptance of contrary views, not actively speaking out on their behalf. Also while those who are politically tolerant accept the rights of groups they abhor to express their views, fanatical groups may feel they have the right to express their point of views though they may be intolerant toward others (Andsager, Wyatt, & Martin, 2004). This study takes its cue from the work by Sniderman and associates (1989) that looks at the degree to which people are willing to support providing a range of rights to their least-liked group, including the right to freedom of expression.

and people are more likely to criticize members of the opposite party (Abramowitz & Saunders, 2005). Political tolerance, then, is central to the argument that selective exposure endangers the political process because those who only seek out opinions from one perspective tend to be intolerant of other perspectives. Some scholars have studied whether exposure to partisan content leads people to become intolerant of opposing perspectives For instance, many researchers who have studied discussion networks have discovered that those who communicate primarily with likeminded individuals are less tolerant of other viewpoints (Hardy, Scheufele & Wang, 2005; Mutz, 2002; Pettigrew, 1997; Pettigrew & Tropp, 2006). However, these studies do not look at selective exposure to the media, but whether discussing politics with those who share your views will make you intolerant of others. Other studies have investigated the influence of tolerance on selective exposure to political Websites and blogs and found that those who are *more* tolerant engaged in selective exposure. The authors speculate that while individuals who are more tolerant may be willing to listen to other perspectives, that does not mean they will seek them out online (Johnson, et al, 2008; 2009). However, these studies focus on the influence of political tolerance on selective exposure rather than selective exposure on tolerance.

Because studies could not be found that have directly explored the influence of selective exposure to online media to levels of political tolerance, this study employs the following research question:

RQ1: What is the relationship between levels of selective exposure to blogs and levels of political tolerance?

FACE-TO-FACE POLITICAL DISCUSSION AND POLITICAL TOLERANCE

Studies suggest that who one has discussions with has a greater influence on political tolerance than amount of face-to-face discussion (Brundidge, 2006; Mutz & Mondak, 2006; Scheufele et al, 2006). Studies suggest that network heterogeneity, that is, discussing politics with people of varying viewpoints, produces several positive effects, including support for democratic outcomes (Price, Cappella & Nir, 2002), the ability to generate reasons for one's political opinions and increased tolerance of those with differing viewpoints (Brundidge, 2006, 2007, 2008; Mutz & Mondak, 2006). Researchers have noted, however, that face-to-face interactions with people of different political perspectives may be decreasing as activities that once required face-to-face interaction can now be done online (Mutz & Mondak, 2006). While diverse political discussions do occur with one's neighbors, friends and associates (Brundidge, 2006, 2008; Scheufele et al., 2006), people are increasingly drawn to residential communities, churches, and volunteer groups that share one's views and lifestyles, thereby limiting exposure to politically diverse points of view (Brundidge, 2006; Huckfeldt & Sprague, 1995; Mutz & Mondak, 2006). Therefore, results suggest that amount of political discussion one engages in is weakly associated with levels of tolerance, but shielding oneself from opposing views is strongly associated with low tolerance. Based on this assertion, this study predicts:

H1a: Reliance on face-to-face discussion for political information will be unrelated to political tolerance.

MEDIA AND POLITICAL TOLERANCE

How the media frame the news can greatly affect political tolerance. Perceived threat has been identified as one of the strongest predictors of political tolerance. For most citizens, perceptions of fear are created by their information environment. If the media portray a group as operating outside the normal boundaries of accepted behavior then citizens, even those predisposed to tolerance, will judge that group harshly (Keum et al., 2005; Sullivan & Transue, 1999). Similarly, studies suggest when individuals perceive their way of life is endangered by an external threat, such as the months after 9/11, increased fear and anger can cause individuals to abandon their standing commitment to democratic values and be willing to sacrifice civil liberties for safety and security, leading to increased intolerance of any groups perceived as threatening (Gibson, 1992; Skitka, Bauman & Mullen, 2004; Marcus, Sullivan, Theiss-Morse & Stevens, 2005; Marcus, Sullivan, Tbeiss-Morse, & Wood, 1995).

BLOG RELIANCE AND POLITICAL TOLERANCE

While balance and fairness are cannons of traditional journalism, the Internet provides a wealth of information targeted to specific viewpoints (Albrecht, 2006) and the Internet's interactivity provides the user more control over what sites to visit (Garrett, 2006). Studies have shown that one way people exercise control of what they are exposed to is by accessing political information that primarily supports their viewpoint (Bimber & Davis, 2003; Garrett, 2006, Mutz, 2001), which tends to lessen political tolerance (Bimber & Davis, 2003; Garrett, 2006; Johnson, et al., 2008, 2009; Mutz & Martin, 2001; Stroud, 2006, 2007, 2008).

While not all political blogs espouse a viewpoint, studies suggest that most political blogs advocate a particular ideology and therefore create a community by attracting likeminded individuals who hold opinions similar to the blogger's (Johnson et al., 2007; Johnson, et al., 2008, 2009; Kaye, 2007; Kaye & Johnson, 2004; Perlmutter, 2008). Because of the highly partisan nature of blogs, they may discourage political tolerance. Johnson, Bichard and Zhang (2009) specifically explored selective exposure to blogs and found that blog users were significantly more likely than political website users to read online content that expresses viewpoints similar to their own, although both blog and political Website reliance are significant predictors of selective exposure (Johnson, Zhang & Bichard, 2008). This finding supports earlier studies that suggest that selective exposure is stronger for media with more partisan content (Brannon, Tagler, M, & Eagly, 2007; Best, Chmielewski, & Kruger, 2005; Stroud, 2006, 2007, 2008). Therefore this study predicts:

H1b: Strong reliance on political blogs will result in lower levels of political tolerance.

POLITICAL MEASURES AND POLITICAL TOLERANCE

Studies going back to the early work of Protho and Grigg (1960) and McCloskey (1964) found that political activists were more likely to support democratic principles such as free speech and minority rights and were more likely to apply them to specific situations. Therefore, activists were the ones that kept democracy functioning (McCloskey, 1964).

Similarly, studies of political expertise claim that those who are more politically interested, psychologically involved, active and knowledgeable are also more politically tolerant. Expertise is linked to higher levels of political sophistication and therefore those with higher expertise report more support of democratic values and they recognize the link between their general value orientations and specific applications of tolerance (Peffley, Knigge, & Hurwitz, 2001). On the other hand, Sullivan and associates (1982) found that while political involvement (a combination of political participation and political knowledge) did increase tolerance, that the relationship was more a function of factors such as level of education and income rather than a greater support for democratic principles than less involved individuals. Therefore, political involvement dropped out as a predictor in their final political tolerance model.

Most studies have found that social liberals are more accepting of minority opinions and of groups who are not part of the mainstream than conservatives (Becker & Scheufele, 2009; Bobo & Licari, 1989; Keum et al., 2005; Linder & Nosek, 2009; Lipset & Raab, 1970). In fact, tolerance is sometimes defined in terms of holding liberal attitudes toward groups outside the mainstream (Lipset & Raab, 1970). Some researchers have questioned whether liberals are any more tolerant of beliefs that they find abhorrent than conservatives (Chong, 2006) or whether both liberals and conservatives are intolerant toward groups on the other ideological extreme. Sullivan and associates (1982) did find that liberals were marginally more tolerant than conservatives, but that neither group is particularly tolerant of views on the opposite end of the ideological spectrum.

H2a: Higher levels of online political participation will predict higher levels of political tolerance.

H2b: Higher levels of offline political participation will predict higher levels of political tolerance.

H2c: Higher levels of political knowledge will predict higher levels of political tolerance.

H2d: Liberals will report higher levels of political tolerance than conservatives.

Demographics and Political Tolerance

Demographic measures, particularly age and socioeconomic factors, have typically emerged as significant predictors of political tolerance (Andsager, Wyatt & Martin, 2004; Bobo & Licari, 1989; Wilson, 1994), although some studies have found few differences (Sullivan et al, 1982). Those who are more educated and have higher incomes are more politically tolerant because creating a consistent position on democratic values and being able to apply democratic values to concrete situations requires increased cognitive skills. Education is also linked to increased openness to new information that could lead to increased tolerance (Bobo & Licari, 1989). Finally, schools teach students the importance of a free marketplace of ideas being important to democracy and attending school puts people in contact with those who have different ideas from one's own (Sullivan et al., 1982).

Researchers also argue that political positions often harden over time so that the young report more political tolerance than older individuals (Stouffer, 1955). Andsager and associates (2004) also found that younger people supported both free speech and media rights more than older individuals, but attributed the difference to the fact that younger individuals believe they have more perceived power in influencing how society functions and what government does.

Others argue that generational differences may better explain differences in tolerance. For instance, those coming of age in the 1950s are reportedly low in tolerance because of the rise of communism while those who came of age during the civil rights struggle for women, blacks and the poor expressed more political tolerance (Nunn et al., 1978). On the other hand, other studies suggest that while generational differences in tolerance exist, that overall people became less tolerant as they grew older (Sullivan et al, 1982; Wilson, 1994).

Studies have traditionally found that men are more politically tolerant than women (Bobo and Licari, 1989; Marcus et al., 1995; Nunn et al., 1978). These studies suggests that women are less tolerant than men because of weaker commitment to abstract norms of democracy, which is partially due to less political expertise (Sniderman 1976; Golebiowska, 1999), and because women report less psychological involvement in politics and therefore less political knowledge (Delli Carpini & Keeter, 1999). Women have been found to be less willing to extend rights to those outside of the political mainstream because they perceive these groups pose a greater threat and their greater commitment to traditional morals (Golebiowska, 1999). But as women's levels of political expertise, knowledge and involvement have increased, differences between men's and women's political positions have narrowed (Sotello, 1999). Indeed, Sullivan and associates (1982) find no significant differences based on gender.

H3a: People with higher levels of education will be more politically tolerant than those with lower levels of education.

H3b: People with higher incomes will be more politically tolerant than those with lower incomes.

H3c: Young people will be more politically tolerant than older individuals.

RQ2: Will women or men report higher levels of political tolerance?

METHODS

Data Collection

An online survey investigating the practice of selective exposure was posted from September 13 to September 23, 2007. Politically interested Internet users were solicited from an online panel operated by a media research lab at a major university. E-mail invitations were sent to 5,000 panel members encouraging their participation if they used the Internet to search for political information and were eligible to vote in the United States. Respondents were also offered an incentive in the form of a cash drawing. The sample yielded 772 completed surveys, a 15.4% response rate.

MEASURES

Demographic Variables

Almost 60 percent (59.8%) of the sample respondents were female. The median age was 44 years old (SD = 12.7). On average, respondents reported obtaining a four-year college degree (SD = 1.1). These variables closely resembled the panel demographics as provided by the research lab used in this study.[2] Respondents also reported a median household income between \$65,001 and \$80,000. Also, 80.8% of the respondents were whites, 7.2% Hispanics, 4.6% blacks, 3.8% Asians, and .5% Native Americans. The race variable was recoded as a dummy variable where white was "1" and nonwhites as "0."

Political Variables

Strength of party affiliation was measured with an 11-point scale where "1" meant "very weak party affiliation" and "11" meant "very strong party affiliation," respondents showed fairly strong party affiliation with an average of 6.8 (SD = 2.7).

Also, 41.2% considered themselves moderate, 21.8% conservative, and 22.5% liberal. In addition, 31.6% identified themselves as Republicans, 36.4% Democrats, 27.6% Independents, with the remainder responding "Libertarian" or "Green Party."

General political interest was measured on an 11-point scale ranging from "1" "absolutely not interested in politics" to "11" "absolutely interested in politics." Respondents are fairly interested in politics in general (M = 7.6, SD = 2.5).

Offline political participation was measured by an index of five items adapted from the National Election Studies where "1" meant "never" participated and "5" meant "very often." Respondents were asked how often they had engaged in a variety of political activities (such as wearing a campaign button, helping or giving money to a candidate, attending political meetings, and persuading others to support a candidate) during the past two years (Cronbach's alpha = .88).

On the same 5-point scale as offline political participation, *online political participation* was an additive measure of seven items taken from the Pew Research Center. Respondents were asked how often they participated in any of the following activities when they went online: (1) registering their own opinions by participating in an online poll, (2) getting information about a candidate's voting record, (3) sending or receiving emails about the candidates or campaigns, (4) contributing money online to a candidate running for public office, (5) looking for more information online about candidates' positions on the issues, (6) finding out about endorsements or ratings of candidates by organizations or individuals

---

2 For gender, the research lab panel demographics were 39.9% male and 60.1% female. Similarly, our sample yielded 40.2% male and 59.8% female. The Media Research Lab panel age breakdown given was 18-24 (9.6%), 25-34 (34%), 35-44 (26.2%), 45-54 (17.8%), 55+ (12.3%) while our sample yielded 18-24 (.1%), 25-34 (25%), 35-44 (25.3%), 45-54 (24.9%), 55+ (24.7%). The education profile for the majority of respondents on the Media Research Lab panel indicated some college up to a four -ear degree (58.5%). The majority of our sample also reported some college up to a four-year degree (57.5%). Therefore, our sample closely resembled the population on gender and amount of education, but is older the panel population as a whole.

online, and (7) checking the accuracy of claims made by or about the candidates online (Cronbach's alpha = .89).

Political knowledge: Respondents were asked four questions taken from the Pew Research Center: if Americans bought more foreign goods than they sold to people overseas, or sold more goods than they bought, or about the same; if they know which political party has a majority in the U.S. Congress; what was the new minimum wage; whether U.S. Supreme Court Chief Justice is more liberal or conservative, and if in 2006, more Iraqi civilians or more U.S. soldiers were killed as a result of the fighting in Iraq. Correct answers were recoded as "1" and incorrect ones as "0" so that the higher the number the more knowledgeable respondents were.

Reliance of Political Blogs for Political Information

Respondents were the extent to which they relied on political blogs for political information on a 5-point scale where "1" meant "don't rely on at all" and "5" meant "heavily rely on": *blogs* (M = 1.71, SD = .99).

Reliance of Offline Discussion

Offline Discussion: This was a single item measure. Respondents were asked to what extent they rely on face-to-face discussion with others for their political information on a 5-point scale where "1" meant "don't rely on at all" and "5" meant "heavily rely on."

Selective Exposure

Respondents used a 5-point scale to indicate the frequency of visits to political weblogs that provide information about specific issues or policies that they agree with and disagree with ranging from "1" "never visited" to "5" "very often visited." Responses for "disagree with" were reverse coded. The two items were summed to form an index of *selective exposure to political weblogs* (high number means high selective exposure).

Political Tolerance

This was an additive measure of six items taken from Sniderman and associates (1989). Respondents were asked to nominate their least-liked group and were asked if their least-liked group should be allowed to exist and to enjoy certain civil rights (be allowed to run for public office, teach in universities, make a speech, have books in public libraries, and be able to hold public rallies in one's city) based on "yes" and "no" answers. Answers were recoded so that "1" meant "high tolerance" and "0" "low tolerance." Then the six items were summed to form a political tolerance index (Cronbach's alpha = .76).

DATA ANALYSIS

Path analysis techniques were used to examine the direct and indirect paths to political tolerance, to shed light on the tenability of the causal models, and to test the hypotheses in the current study. To obtain the path coefficients, various regression analyses were conducted. The first regression analysis employed political tolerance as the dependent variable, with the following four blocks of independent variables entered in this order: block 1: selective exposure to blogs; block 2: reliance on political blogs and reliance on face-to-face discussion; block 3, political variables (political interest, offline and online political participation, political knowledge, strength of party affiliation, and ideology); block 4: demographics (gender, age, education, income, and race).

Selective exposure to political blogs was the dependent variable for the second regression analysis. It was regressed on the following three blocks of independent variables that were entered in this order: block 1: reliance on political blogs and reliance on face-to-face discussion; block 2, political variables (political interest, offline and online political participation, political knowledge, strength of party affiliation, and ideology); block 3: demographics (gender, age, education, income, and race).

The fourth regression analysis had the reliance on politically oriented blogs as the dependent variable. The following two blocks of independent variables were entered in this order: block 1, political variables (political interest, offline and online political participation, political knowledge, willingness to speak in public on controversial topics, strength of party affiliation, and ideology); block 2: demographics (gender, age, education, income, and race).

The fifth and sixth regression analyses had the reliance on political blogs and reliance on face-to-face discussion as the dependent variable respectively. The following two blocks of independent variables were entered in this order: block 1, political variables (political interest, offline and online political participation, political knowledge, strength of party affiliation, and ideology); block 2: demographics (gender, age, education, income, and race).

The path coefficients are shown in Tables 1-4.

Table 1. Effects of Selective Exposure to Political Blogs on Political Tolerance

	Direct Effects	Indirect Effects	Total Effects
Selective exposure to political blogs	.03	0	.03

Note:* = p <.05, ** = p < .01, *** = p<.001

Percentage of variance accounted for by block of selective exposure variables: .6*

Table 2. Effects of Reliance on Political Blogs and Reliance on Face-to-Face Discussion on Political Tolerance

	Direct Effects	Indirect Effects	Total Effects
Reliance on political blogs	-.04	.01	-.03
Reliance on face-to-face discussion	.09*	0	.09*

Note:* = p <.05, ** = p < .01, *** = p<.001

Percentage of variance accounted for by block of variables of reliance on blogs, and face-to-face discussion: 2.4*

Table 3. Effects of Political Variables on Political Tolerance

	Direct Effects	Indirect Effects	Total Effects
Political knowledge	.18***	0	.18***
Ideology	.09*	0	.09*
Offline political participation	0	.02**	.02**
Online political participation	0	.02**	.02**
Strength of party affiliation	0	-.07*	-.07*
Political interest	0	.01*	.01*

Note:* = p <.05, ** = p < .01, *** = p<.001

Percentage of variance accounted for by block of political variables: 9.6**

Table 4. Effects of Demographic Variables on Political Tolerance

	Direct Effects	Indirect Effects	Total Effects
Education	.14**	0	.14**
Income	.10*	0	.10*
Age	-.02	-.01*	-.01*
Gender	-.04	0	-.04
Race	.07	0	.07

Note:* = p <.05, ** = p < .01, *** = p<.001

Percentage of variance accounted for by block of demographic variables: 3.9**

RESULTS

Effects of Selective Exposure on Political Tolerance

RQ1 questions the impact of selective exposure to blogs on political tolerance. According to Table 1, selective exposure to political blogs has no significant effect on political tolerance ($\beta = .03$, n.s.). The block of selective exposure variables account for .6% of the variance in political tolerance.

Effects of Reliance on Political Blogs and Face-to-Face Discussion on Political Tolerance

H1a predicts that reliance on face-to-face discussion will be unrelated to political tolerance. As seen from Table 2, reliance on face-to-face discussion has a direct effect on political tolerance ($\beta = .09$, $p<.05$), so H1a is not supported. H1b predicts that reliance on political blogs leads to less tolerance. As seen from Table 2, reliance on political blogs has no effect on tolerance, either ($\beta = -.03$, n.s.). So, H1b is not supported. The block of reliance on blogs and face-to-face discussion account for 2.4% of the variance in political tolerance.

Effects of Political Variables on Political Tolerance

H2a predicts that people who participate more online in politics are more tolerant. According to Table 3, though there is no direct link between online political participation and tolerance, online political participation influences reliance on face-to-face political discussion (β = .20, p<.001), which, in turn, influences political tolerance (β = .09, p<.05) with an indirect of .02 (p<.01). Thus, H2a is partially supported.

H2b states that people who participate more offline in politics are more tolerant. Table 3 shows that offline political participation has no direct effect on tolerance.

However, offline political participation influences reliance on face-to-face political discussion (β= .20, p<.001), which, in turn, influences political tolerance (β = .09, p<.05) with an indirect of .02 (p<.01). Thus, H2b is partially supported.

H2c states that people with more political knowledge are more tolerant. As seen from Table 3, political knowledge has a direct effect on tolerance (β = .18, p<.0001), so H2c is supported.

H2d states that liberals are more tolerant than conservatives. Table 3 shows that ideology (liberal coded higher) has a direct effect on tolerance (β = .09, p<.05). Thus, H2d is supported.

Table 3 also shows that strength of party affiliation influences tolerance via reliance on face-to-face discussion (β = -.01, p<.05), meaning strong partisans are less tolerant. Political interest has a positive effect on tolerance via reliance on face-to-face discussion (β = .01, p<.05).

The block of political variables accounts for 9.6% of the variance in political tolerance, by far the largest of any block.

Effects of Demographic Variables on Political Tolerance

H3a posits that people with more education are more tolerant. Table 4 shows that education has a direct effect on tolerance (β = .14, p<.01), so H3a is supported.

H3b predicts that people with higher income levels are more tolerant. Income also has a direct effect on tolerance (β = .10, p<.05), so H3b is supported (Table 4).

H3c states that young people are more tolerant. As seen from Table 4, age does not have a direct effect on political tolerance, but it influences reliance on face-to-face discussion (β = -.08, p<.05), which in turn, influences political tolerance (β = .09, p<.05), with an indirect effect of -.01, p<.05. Therefore, H3c is partially supported.

RQ2 posits the question of whether gender will have an impact on reported levels of political tolerance. According to Table 4, gender has no effect on political tolerance (β = .07, n.s.).

The block of demographic variables account for 3.9% of the variance in political tolerance.

Discussion

The current study provided a unique opportunity to address the important issue of political tolerance as it relates to selective exposure. Through an online survey, respondents were questioned regarding their levels of selective exposure to political blogs. This paper defines selective exposure in the classic sense; the practice of seeking information that supports one's point of view and avoiding information that is inconsistent with one's convictions. This research also measures reliance on blogs and face-to-face discussion, political attitudes and behaviors, along with demographic variables to paint an accurate portrait of the factors most likely to impact political tolerance.

The findings indicate that selective exposure to political blogs has no effect on political tolerance. While this study revealed no relationship, past analyses of selective exposure have indicated mixed results. Previous research has noted that polarized opinions lead to less tolerance of opposing viewpoints (Mutz, 2002). Past studies have also found that those who are less tolerant are more likely to exercise selective exposure (Hardy, Scheufele & Wang, 2005; Mutz, 2002; Pettigrew, 1997; Pettigrew & Tropp, 2006). Other analyses assert that those who are more tolerant are also more likely to selectively expose themselves to political information (Johnson, Zhang & Bichard, 2008). This inconsistency in findings may be due to the specific nature of the analysis. The current study addressed the impact of selective exposure solely to blogs. Past research has often relied upon a general Internet exposure variable, which may contribute to the difference in results.

When analyzing measures of reliance, this analysis hypothesized that the *amount* of face-to-face discussion would be unrelated to political tolerance. This hypothesis was not supported as face-to-face discussion emerged as the only reliance measure to directly influence political tolerance. This finding is somewhat surprising because scholars have found that the *amount* of face-to-face discussion as a factor leading to political tolerance pales in comparison to *who* one has discussions with (Brundidge, 2006; Mutz & Mondak, 2006; Scheufele et al, 2006). While discussions with those holding a variety of viewpoints may align with increased tolerance, discussions with strictly homogeneous groups would not (Price, Capella & Nir, 2002). It may be that our respondent pool relies heavily on a heterogeneous network for communication and that is why face-to-face discussion impacted political tolerance positively. Future examinations will need to address not only the amount of discussion offline, but with *whom* those discussions take place in order to reveal more information about this relationship.

Reliance on political blogs appeared to have no impact on reported levels of political tolerance. This finding contradicts past analyses that indicate reliance on online media (especially the highly partisan content of blogs) is related to the development of polarized opinions and political intolerance (Bimber & Davis, 2003; Mutz, 2006; Johnson, Bichard & Zhang, 2008; Kaye & Johnson, 2004). Possibly, individuals that rely heavily on select online media sources are able to do so without compromising their ability to listen and interact with those holding opposing viewpoints.

As hypothesized, political variables were very effective predictors of political tolerance. Increased political knowledge directly predicted greater political tolerance. This result supports past analyses that have found political sophistication linked with increased tolerance (Peffley, Knigge & Hurwitz, 2001). Online and offline political participation indirectly led to increased political tolerance among participants. Both of these participation variables influenced reliance on face-to-face discussion, which, in turn, impacted political tolerance. It would seem as though individuals that are more engaged with the political process both in their offline and online spheres come into contact with, and appreciate, diverse viewpoints. Findings also surfaced indicating that ideological differences play a role in one's political tolerance. As noted in past analyses (Bobo & Licari, 1989; Keum et al, 2005; Lipset & Raab, 1970), liberals reported higher levels of political tolerance when compared to conservatives.

The results demonstrate that various demographic measures, in fact, have an impact on political tolerance. Those who are more educated and report higher incomes are more politically tolerant. This aligns with past research asserting such measures indicate higher cognitive ability and therefore more openness to opposing views (Bobo & Licari, 1989). Younger participants also appear to exert more tolerance, although the results indicate that this as an indirect connection by way of increased reliance on face-to-face discussion. The question of gender effects on political tolerance revealed no relationship. This is interesting because a great deal of prior analyses have noted that women are likely to be less tolerant because of decreased levels of political expertise and involvement (Bobo & Licari, 1989; Golebiowska, 1999; Delli, Carpini & Keeter, 1999).

Ultimately, this research revealed a very telling profile of politically tolerant individuals. The findings denote that those who are young, rely heavily on face-to-face political discussion, are politically knowledgeable and active (both online and off), espouse a more liberal ideology, are well educated and report higher incomes demonstrate more political tolerance. Face-to-face discussion was revealed as a dominant factor in many of the relationships explored in the current study. Whether directly or indirectly, face-to-face political interaction appears to help enhance political tolerance. Perhaps the personal quality inherent with face-to-face discussion makes it more difficult to react with intolerance. Communication online can be anonymous, whereas offline discussion requires that you attach an actual face with an opinion.

The study of political tolerance and those who display it is one worthy of continuation. American democracy is founded on the ability to freely express political views and opinions in an environment that is conducive to such expression. Tolerance of differences in opinion provides a rich atmosphere for democratic stability and growth. The investigation into what contributes to such tolerance is important to the preservation of civil liberties and freedom of expression.

LIMITATIONS

Data collection for the current study relied solely on survey methods. This is a limiting factor in that survey questionnaires provide a simple one-time glimpse of respondent opinions. Future research should either look at the effect of selective exposure on tolerance over time or employ an experiment to better determine cause and effect between selective exposure and tolerance. As always, a larger and more diverse profile of participants would also enrich the results. Future analyses should also take the opportunity to investigate political tolerance and related measures in the context of a political election. Future studies may also want to include other Internet components such as political websites and social media.

REFERENCES

Abramowitz, A., & Saunders, K. (2005). Why can't we all just get along? The reality of a polarized America. *The Forum*: Vol. 3 : Iss. 2, Article 1. Retrived May 6, 2009 from http://www.bepress.com/forum/vol3/iss2/art1/.

Albrecht, S. (2006). Whose voice is heard in online deliberation? A study of participation and representation in political debates on the Internet. *Information, Communication & Society*, 9, 62-82.

Andsager, J.L, Wyatt, R.O., & Martin, E. (2004). *Free expression in 5 democratic publics*. Creskill, NJ: Hampton Press.

Baldassarri, D., & Bearman, P. (2007). Dynamics of political polarization. *American Sociological* Review, 72, 784-811.

Becker, A.B., & Scheufele, D.A. (2009). Moral politicking: Public attitudes toward gay marriage in an election context. *The International Journal of Press/Politics, 14*, 186-211.

Best, S. J., Chmielewski, B., & Krueger, B. S. (2005). Selective exposure to online foreign news during the conflict with Iraq. *Harvard Journal of Press/Politics, 10*(4), 52-70.

Bibo, L., & Licari, F.C. (1989). Education and political tolerance: Testing the effects of cognitive sophistication and target group affect. *Public Opinion Quarterly, 53*, 285-308.

Bimber, B., & Davis, R. (2003). *Campaigning online: The Internet in U.S. elections*. Oxford: Oxford University Press.

Brannon, L.A., Tagler, M.J., & Eagly, A.H. (2007). The moderating role of attitude strength in selective exposure to information. *Journal of Experimental Social Psychology* 43, 611-617.

Brundidge, J. (2006, June). *The contribution of the Internet to the heterogeneity of political discussion networks: Does the medium matter?* Paper presented at the International Communication Association Conference, Dresden, Germany.

Brundidge, J. (2007, May). *Modeling the impact of the Internet on exposure to disagreement*. Paper presented at the International Communication Association Conference, SanFrancisco, CA.

Brundidge, J. (2008, May). *The contemporary media environment and breadth of communication: The contribution of the Internet to the heterogeneity of political discussion networks*. Paper presented at the International Communication Association Conference, Montreal, CN

Chong, D. (2006). Free speech and multiculturalism in and out of the academy. *Political Psychology, 27*, 29-54.

Delli Carpini, M. X., & Keeter, S. (1996). *What Americans know about politics and why it matters*. New Haven: Yale University Press

Etzioni, A. (2002). Are particularistic obligations justified? A communitarian examination. *The Review of Politics, 64*(4): 573-598

Fiorina, M.P., Abrams, S.J., & Pope, J.C. (2005). *Cultural wars? The myth of polarized America*. New York: Pearson Longman.

Garrett, R.K. (2006). Protest in an information society: A review of literature on social movements and new ICTs. *Information, Communication and Society*, 9, 202-224.

Gibson, J.L. (1992). Alternative measures of political tolerance: Must tolerance be "least-liked"? *American Journal of Political Science, 36,* 560-577.

Gibson, J.L. (1986). Political intolerance in America: A reconsideration. *American Politics Quarterly, 14,* 267-293.

Golebiowska. E.A. (1999). Gender gap in political tolerance. *Political Behavior, 21,* 43-6

Graf, J., & Aday, S. (2008). Selective attention to online political information. *Journal of Broadcasting & Electronic Media,* 52, 86-100.

Hardy, B. W., Scheufele, D. A., & Wang, Z. (2005, May). *Defining deliberation: Key determinants and distinct dimensions.* Paper presented to the annual convention of the International Communication Association. New York, NY.

Huddy, L., Khatid, N., & Capelos, T. (2002). Trends: Reactions to the terrorism attacks of September 11, 2001. *Public Opinion Quarterly*, 66, 418-450.

Huckfeldt, R., & Sprague, J. (1995). *Citizens, politics and social communication.* Cambridge: Cambridge University Press.

Iyengar, S., & Hahn, K. (2009). Red media, blue media: Evidence of ideological selectivity in media use. *Journal of Communication, 59*, 19-39.

Johnson, T. J., Bichard, S.L., & Zhang, W. (2009). Communication communities or 'cyberghettos?: A path analysis model examining factors that explain selective exposure to blogs. *Journal of Computer-Mediated Communication, 15,* 60-82.

Johnson, T. J., Bichard, S. L., & Zhang, W. (2008, May*). Revived and refreshed: Selective exposure to blogs and political Websites.* Paper presented at the International Communication Association annual convention, Montreal.

Johnson, T. J. & Kaye, B. K. (2004). Wag the Blog: How reliance on traditional media and the Internet influence perceptions of credibility of Weblogs among blog users. *Journalism & Mass Communication Quarterly,* 81 (3), 622-642.

Johnson, T. J. & Kaye, B. K. (2006). Blog day afternoon: Are blogs stealing audiences away from traditional media sources? in Ralph Berenger (ed.), *Cybermedia go to war* (316-333). Spokane, WA: Marquette Books.

Johnson, T. J., Kaye, B.K., Bichard, S. L., & Wong, W. J. (2007). Every blog has its day: Politically interested Internet users' perceptions of blog credibility. *Journal of Computer-Mediated Communication,* 13, article 6. http://jcmc.indiana.edu/vol13/issue1/johnson.html.

Johnson, T.J., Zhang, W., & Bichard, S.L, (2008, August). *Voices of convergence orconflict?: A path analysis investigation of selective exposure to political websites.* Paper submitted to the Association for Education in Journalism and Mass Communication annual convention, Chicago,IL.

Kaye, B. K. (2007). Blog use motivations: An exploratory study. In M. Tremayne (ed.). *Blogging, citizenship, and the future of media.* (pp. 127-148). New York: Routledge.

Kaye, B. K., & Johnson, T. J. (2004). Weblogs as a source of information about the 2003 Iraq War. In R.D. Berenger (ed), *Global media goes to war: Role of news and entertainment media during the 2003 Iraq War* (pp. 291-301). Spokane, WA: Marquette Books.

Keum, H., Hillback, E.D., Rojas, H., De Zuniga H.G., Shah, D.V., Mcleod, D.M. (2006). Personifying the radical: How news framing polarizes security concerns and tolerance judgments. *Human Communication Research*, 31, 337-364

Knobloch-Westerwick, S., & Meng, J. (2009). Looking the other way: Selective exposure to attitude-consistent and counter-attitudinal political information. *Communication Research*, 36, 426-448.

Linder, N.M., & Nosek, B.A. (2009). Alienable speech: Ideological variations in the application of free-speech principles. *Political Psychology, 30,* 67-92.

Lipset, S.M., & Raab, E. (1970*). The politics of unreason: Right-wing extremism in America, 1790-1970.* New York: Harper & Row, Publishers.

Marcus, G.E., Sullivan, J.L, Theiss-Morse, E. Stevens, D. (2005). The emotional foundation of political cognition: The impact of extrinsic anxiety on the formation of political tolerance judgments. *Political Psychology, 26,* 949-963.

Marcus, G.E., Sullivan, J.L., Theiss-Morse, E., & Wood, S.L (1995*). With malice toward some:How people make civil liberties judgments.* Cambridge: Cambridge University Press.

McCloskey, H. (1964). Consensus and ideology in American politics. *American Political Science Review*, 58, 361-382.

Meffert, M. F., Chung, S., Joiner, A. J., Waks, L., & Garst, J. (2006). The effects of negativity and motivated information processing during a political campaign. *Journal of Communication, 56*(1), 27-51.

Mueller, J. (1988). Trends in political tolerance. *Public Opinion Quarterly*, 52, 1-25.

Mutz, D. C. (2002). Cross-cutting social networks: Testing democratic theory in practice. *American Political Science Review, 96*(1), 111-126.

Mutz, D.C. (2006). *Hearing the other side: Deliberative versus participatory democracy.* Cambridge: Cambridge University Press.

Mutz, D. C., & Martin, P. S. (2001). Facilitating communication across lines of political difference: The role of mass media. *American Political Science Review, 95*(1),

Mutz, D. C. & Mondak, J.J. (2006.) The workplace as a context for cross-cutting political discourse. *The Journal of Politics, 68*, 140-155

Nunn, C.Z., Crockett Jr., H.J., and Williams, J.A. (1978). *Tolerance for nonconformity.* San Francisco: Jossey-Bass.

Peffley, M. Knigge, P., & Hurwitz, J. (2001). A multiple values model of political tolerance. *Political Research Quarterly, 54*, 379-406

Perlmutter, D. A. (2008). *Blog wars.* New York: Oxford University Press.

Pettigrew, T.F. (1997). Generalized intergroup contact effects on prejudice. *Personality and Social Psychology Bulletin* 23, 173-185.

Pettigrew T.F., and Tropp, L. (2006). A meta-analytic test of intergroup contact theory. *Journal of Personality and Social Psychology, 90*, 751-783.

Price, V., Capella, J.N. & Nir, L. (2002). Does disagreement contribute to more deliberative opinion? *Political Communication, 19,* 95–112.

Protho, J.W., & Grigg, C.M. (1960). Fundamental principles of democracy: Bases of agreement and disagreement. *Journal of Politics, 22*, 276-294.

Scheufele, D.A., Hardy, B.W., Brossard, D., Waismel-Manor, I.S., & Nisbet, E. (2006). Democracy based on difference: Examining the links between structural heterogeneity, heterogeneity of discussion networks, and democratic citizenship. *Journal of Communication* 56, 728-753.

Scheufele, D.A., Nisbit, M.C., Brossard, D., & Nisbet, E.C. (2004) Social structure and citizenship: Examining the impact of social setting, network homogeneity, and information variables on political participation. *Political Communication*, 21, 315-338.

Skitka, L. J., Bauman, C. W., & Mullen, E. (2004). Political tolerance and coming to psychological closure following the September 11, 2001 terrorist attacks: An integrative approach. *Personality and Social Psychology Bulletin,* 30, 743–756.

Sniderman, P.M, Tetlock, P.E., Glaser, J,M, Green, D.P., & and Hout, M. (1989). Principled tolerance and American political values. *British Journal of Political Science*, 19, 25-46.

Sotello, M.J. (1999). Gender differences in political tolerance among adolescents. *Journal of Gender Studies, 8,* 211-217.

Stouffer, S.A. (1955). *Communism, conformity and civil liberties: A cross-section of the nation speaks its mind.* New York: Doubleday and Co.

Stroud, N.J. (2006). *Selective exposure to partisan information.* Unpublished dissertation, University of Pennsylvania.

Stroud, N.J. (2007). Media effects, selective exposure, and *Fahrenheit 9/11. Political Communication, 24,* 415-432

Stroud, N. J. (2008). Media use and political predispositions: Revisiting the concept of selective exposure. *Political Behavior, 30,* 341–366.

Sullivan, J. L., Piereson, J.,& Marcus, G.E. (1982*), Political tolerance and American democracy.* Chicago: The Univesity of Chicago Press.

Sullivan, J.L. & Transue, J.E. (1999). The psychological underpinnings of democracy: A selective review of research on political tolerance, interpersonal trust, and social capital. *Annual Review of Psychology, 50*, 625-650

Sunstein, C. (2001). *Republic.com*. Princeton: Princeton University Press.

Sunstein, C. (2007). *Republic.com 2.0.* Princeton: Princeton University Press.

Wall, M. (2006). Blogging Gulf War II. *Journalism Studies*, 7(1), 111-126.

Wilson, T.C. (2004). Trend in tolerance toward rightist and leftist groups, 1976-1988: Effects of attitude change and cohort succession. *Public Opinion Quarterly, 58*, 539-556

In: Internet Issues: Blogging, the Digital Divide and Digital... ISBN: 978-1-60741-824-5
Editor: Cassie M. Evans, pp. 165-176

Chapter 10

EQUITY, JUSTICE AND THE DIGITAL DIVIDE IN A PANDEMIC

Connal Lee
Flinders University of South Australia

INTRODUCTION

This chapter is concerned with equity and communication in a pandemic and analyses issues surrounding the potential impact of the digital divide during such an emergency. I will look at inequalities related to communication in particular access to information and how the internet can be both a positive and negative force in addressing or exacerbating these inequalities.[1] The methodology here will be to look at the role of communication in a pandemic and establish how the internet factors into debates surrounding pandemic communication.

Communication during a pandemic is a vital consideration; ethical concerns arise regarding the possible harms posed by a pandemic and the ways in which existing inequalities in communication might place people in harms way. My argument is that the digital divide is not only potentially corrosive because access to the internet has some value but the digital divide is as much about inequalities in resources, capabilities and infrastructure. An example of this being inequalities between developed and well-resourced nations and less well-off nations and the impact that this division can have in a global pandemic. Lacking a communication infrastructure, for instance a media or technologies such as the internet is somewhat symptomatic of lacking other types of resources and structures. As such I will sketch the terrain of the issue of equity and communication and highlight the important justice concerns that surround the digital divide in a pandemic.

[1] These three issues were identified in an earlier work; see Lee, Connal and Rogers, Wendy A. Ethics, Pandemic planning and Communications. *Monash Bioethics Review,* Vol. 25 No. 4, October 2006.

1. Pandemic Communication

An initial question is worth asking here; what exactly is a pandemic and what are the potential impacts of a pandemic? Following this we can begin to look at why communication (and issues associated with the internet) is ethically important.

A pandemic by its very nature crosses national borders and can have a potentially wide spread global impact. The effects of a pandemic can be costly to social institutions and human life and lead to a climate of public fear, anxiety and panic. The consequences of a pandemic can be devastating and can threaten society at a variety of levels. As Pierre Saliou points out flu pandemics can constitute "Eminently worrying crisis situations",[2] with the potential to cause harm to society in a variety of ways. Economic fall-out, high rates of morbidity and mortality as well as heavy burdens falling on certain institutions such as those charged with the delivery of health care, are all possible outcomes in a pandemic.[3] Historical cases of infectious disease pandemics have shown that the lives can be lost on a massive scale the 1918 influenza for example is estimated to have been responsible for 25 million deaths.[4]

A pandemic will pose major threats to health and safety and will disrupt normal life in a variety of ways. Measures such as case isolation, household quarantine, school or workplace closure and restrictions on travel are all aspects of pandemic plans which warrant consideration.[5] Typically enjoyed liberties and freedoms to travel and assemble in public will be restricted during a pandemic as means of limiting the spread of contagion. Whilst the effectiveness of contagion limiting arrangements varies with each raising ethical issues the key point here is that effectively meeting a pandemic demands that there are strategies in place for dealing with an infectious disease outbreak. The impact of a pandemic will likely require at least that some measures be put in place which disrupt our normally enjoyed liberties and potentially over-ride rights we might otherwise have.[6] Again ethical issues associated with the extent to which we are justified in moving aside certain constraints to meet emergencies are varied and the key point worth making here is that a pandemic is an emergency scenario where normal life will be disrupted in a potentially very harmful way.

What should we make of some of these features of a pandemic? This is by no means an exhaustive list of the features of a pandemic and at this point I have been talking broadly about an infectious disease pandemic of some sort, as such the particular features of each disease would likely have to be a consideration. My aim thus far has been to draw attention most specifically on the impact of a pandemic involving an infectious disease (with influenza being a prominent example in public health ethics debates). I want to draw attention to the

---

[2] Saliou, Pierre. Crisis communication in the event of flu pandemic. *European Journal of Epidemiology*. 10: 515-517. 1994. p. 516

[3] Smith et al provide a comprehensive analysis of the potential impacts of infectious disease and the specific ethical problems that arise. See; Smith, Charles B. Battin, Margaret P. Jacobson, Jay A, Francis Leslie P, Botkin Jeffrey R, Asplund Emily P, Domek, Gretchen J, and Hawkins, Beverly. Are there characteristics of infectious disease that raise special ethical issues? *Developing World Bioethics*. Volume 4. Number 1. 2004.

[4] See; Hume, Janice. The "Forgotten" 1918 Influenza Epidemic and Press Portrayal of Public Anxiety. *Journalism and Mass Communication Quarterly*; Winter 2000; 77, 4; p. 899. And; Selgelid, Michael. Ethics and Infectious Disease. Bioethics. Volume 19, Number 3. 2005. P. 274

[5] Lee and Rogers. p. 9

[6] This is a large issue that I do not want to explore here however it is worth pointing out that conflicts between individual rights and the public good are an important element of a pandemic response and generally meeting the ends of the public is considered to be morally valuable enough to put normal rules on hold. I want to draw attention here purely to the possible impacts of a pandemic and highlight their seriousness.

disruptive force of a pandemic however the issue at hand here the potential role of communication in mitigating harm(s) and the ethical issues associated with information dissemination in a pandemic. I will now outline how information is a valuable good in a pandemic and can help to limit some of the harms and burdens that a pandemic poses.

2. The Value of Information in an Emergency

Information and the exchange of information can play a role in limiting and mitigating the negative effects of a pandemic. The premise here is uncontroversial; when emergencies strike information can become a vital resource when people are sufficiently informed they are more autonomous they can better protect themselves and they are perhaps less likely to meet certain harms. A good synopsis of what types of information are likely to help mitigate harm(s) is supplied by the Bellagio statement of principles on pandemic influenza who outline how there will be 'people requiring accurate, up-to-date and easily understood information about policy responses and appropriate local and individual actions.[7]

This is of course largely self evident; emergencies threaten life and information provided in a timely fashion can help people to protect themselves. A person trapped in a burning building will be better positioned if they possess information about exists and strategies for escape and survival. On the flip side of this is the impact that little or no information can have on creating anxiety, fear and panic. Lack of knowledge about possible ways of escaping a burning a room would for example be very likely to contribute to feelings of doom and inevitable demise. This is certainly an issue however it also cuts deeper than simply delivering timely accurate messages which enable people to protect themselves better the issue is also about the communications infrastructure of a society and the ways that information is disseminated. Whilst the communications infrastructure of well-developed nations encompasses a variety of forms of media and new technologies it remains that some communities lack any real communication infrastructure and as such any real machinery for disseminating information in an emergency.

To illustrate the potentially corrosive impact that a lack of communication and information can have take this example offered by Laurie Garret which draws attention to harmful impacts that a lack of information in the public domain can have during cases of infectious disease emergencies. Garret uses an example of how public health was compromised in a situation where there was an absence of local media and as such communication in general. 'The 1995 Ebola virus outbreak in Kikwit, Zaire, is a perfect example of what happens in a community facing catastrophe without any media'[8]. Because there were no reliable sources of information locally (no media) all claims were considered equally credible, and doctors came under increasing suspicion[9]. Information spread by word of mouth for instance lead to rumour and here say influencing people's decisions. The World Health Organisation (WHO) intervened, recruiting local medical students to act as 'a media' distributing information and answering questions about the epidemic. Intervention by the

[7] Meeting on Social Justice and Influenza. Bellagio Group Statement of Principles. (2007) available from; http://www.hopkinsmedicine.org/bioethics/bellagio (accessed 25.11.07) p. 1

[8] Garret, Laurie. 'Understanding Media's Response to Epidemics', *Public Health Reports.* 116 (2): 2001: p. 90

[9] Garret, p. 90

WHO helped to mitigate the crisis and once a sustained level of accurate reliable information was achieved so too was the population involved better protected.

These types of considerations draw attention to the role that information plays in a crisis and equally the role that rumour and hearsay can play in placing people unnecessarily in harms way. Garrets example also draws attention to the impact that a lack of communication can have on impoverished and less well-off communities who in the event of a pandemic can become profoundly vulnerable. The concern is as Thomas Glass claims that 'rumours will fill the information gap'.[10] When there is no reliable apparatus for ensuring that important and significant information reaches people this is precisely what can happen and then concern is that people may meet with harm as a result.

The harms posed by a pandemic are first and foremost the disease itself and there appears to be a coherent association between having access to accurate, meaningful and reliable information and being less vulnerable or susceptible to infection. Certainly it is plausible that good effective communication can mitigate harm and this is precisely the rationale that the WHO was working from in the above example. The above case also shows that having a communications infrastructure is necessary for disseminating information. Likewise this case shows that an absence of communication (and a machinery for distributing information) can be intrinsically linked to lacking other important resources. Less well off nations and communities who lack a communications network (such as a media) are also likely to lack other important resources. As such there is a possible link between lacking the means to communicate effectively and being less well off in terms of other important social goods.

The issue here is that a lack of communication is harmful; in a pandemic there needs to be information which is accessible and this cannot happen without some infrastructure. Of course the information needs to meet some conditions in order to be of value, i.e. rumours and here say are not example of valuable information.

I have drawn attention to the important role of having a communications infrastructure, and the potentially negative impacts of lacking such a framework however it cannot be said to be sufficient that we have lots of information floating around the public realm through television media, newspapers or from sources which lack creditability. Organisations such as the media present their own dilemmas and problems too from what they find newsworthy in catastrophes to how well they report accurate and reliable information. Clearly it is a disadvantage to have no information at all however it might be a similar problem if there are too many avenues, sources and outlets for information. More needs to be said about the quality of communication; if we know that having a communication strategy is appropriate for preventing certain harms and if we have a communications infrastructure (for example a media and certain technologies such as the internet) what are the appropriate considerations that need to be made?

A key ethical concern is the capacity for existing inequalities to become exacerbated or simply open people up to harm. In the context of communication we might be concerned about what harms inequalities in access to information open people up to and how inequalities can broaden when communication needs are not met. However I want to say a bit more about the quality of communication and how the internet factors into this discussion before

[10] Glass, Thomas A. 'Understanding Public Response to Disasters'. *Public Health Reports.* Supplement 2. Volume 116. 2001. p. 72

returning to the idea that existing inequalities can broaden in a pandemic and discussing the possible roles of communication in addressing such concerns.

3. THE INTERNET

I now turn to a discussion of the internet which has become a powerful communicative force in contemporary society. On the face of it the internet is a potentially worthwhile ally in emergency communication contributing to making information more accessible. The internet can increase equity in access to information by delivering messages in multiple languages, and by addressing mental and physical barriers to information through tailoring the form and delivery if information.[11] In a previous work looking at these issues national pandemic plans from 12 nations were analysed precisely in terms of how communication is anticipated to play a role in effectively and for that matter ethically responding to pandemic influenza (PI).[12] This study looked not only at how pandemic planners are utilizing communicative tools such as the media and the internet but also at wider ethical issues regarding equity and social justice. In relation to the internet empirical data indicates that policy makers and pandemic planners anticipate that the internet will perform a variety of functions in the delivery and dissemination of vital information.

However, a significant aspect of ethical discourse surrounding the internet is the capacity of such a resource to cut in both a positive and negative fashion; as Deborah Johnson points out, the internet facilitates the spread of misinformation as well as information[13]. The ethical concern is that the internet can, through public health and government websites, provide access to important information about pandemic plans and policy actions however it can also though other social networking sites become a minefield of rumour, hearsay and misinformation. The rise of social networking sites such a Twitter, Facebook and MySpace has fuelled debate about the overall value of the internet as a mechanism for distributing reliable and accurate information. These sites are generally utilised by users for communication which is not constrained by such principles, however their popularity points to a force in modern day communications which cannot (and should not) be ignored.

Pandemic plans from various nations across the world indicate that the internet will feature as a mechanism for distributing information however as the above concern highlights it is not merely though health and authoritative websites that communication will take place. The virtues of government websites are clear; professionals working in the field of pandemic influenza can publish information on such sites which can be utilised by users to make educated decisions and better understand the situation. The virtues of social networking sites are far less clear; users on these sites are not concerned with the dissemination of accurate or reliable messages. Largely this is because such sites are not designed not used with these principles in mind. The ethical tear here can be seen as being between sites that are useful and sites that are problematic.

---

[11] Lee, Connal Rogers Wendy A and Branauck-Mayer, Annette. Social Justice and Pandemic Influenza Planning: The Role of Communications Strategies. *Public Health Ethics*.Volume 1. Number 3. November 2008. p. 229-230

[12] See Lee et al

[13] Johnson, Deborah. *Democratic values and the internet*. From; Langford, Duncan (editor) *Internet Ethics*. MacMillian Press. Hampshire and London. 2000. p. 187

However, the role of the internet in pandemic communications goes beyond consideration of how certain websites promote communication which may prove unhelpful in a pandemic; some sections of the community are not computer literate or internet savvy and such groups may be vulnerable in a pandemic if they cannot access vital messages or directives. The aged, the homeless, recent immigrants, rural dwellers and the economically less well-off could all be said to constitute groups who in the event of a pandemic may be vulnerable to a lack of information.[14]

Putting the issue of problematic websites aside I want to look now at specific inequalities in access to information which have the potential to cause harms in a pandemic and how the internet has the potential to rectify and address these inequalities.

4. Inequalities in Communication

People vary in their ability to both access and act on information. The internet can do some things to improve information access for certain groups however other important inequalities will likely remain. In this discussion I will look at what inequalities in communication are and how the internet can address them.

1. Barriers in Access to Information

This issue is reasonably straightforward; there are people with access to the internet and useful government and public health websites and there are people who lack access to this resource. We might be justifiably concerned that those lacking the resource are missing out on something important. Likewise we might be reasonably worried that those with the resource have access to a variety of information some of which may be incredibly useful and some of which may be potentially harmful. Putting aside the latter category of ethical problems (the spread of poor quality information) the issue of 'the digital divide' and the inequalities that exist between the haves and have nots raises its head. What does it mean to have access to the internet? On the one hand those lacking the internet are being deprived information that may need on the other they are avoiding the rumour mill. Those who do have the resource have access to potentially valuable information however they are also exposed to information and communication which is problematic such as through social networking sites.

The internet can be seen as a useful tool in so far as it reduces inequalities in access to information. In moderately affluent societies without huge disparities in wealth and income it has become the norm for most households to have access to computer technology and the internet. Cass Sunstein for example claims that technological developments are likely to be a great ally to small groups and minorities, however defined, for the very reason that it is easier to access information.[15]

However, it is not as clear cut as Sunstein might claim. One example is that of global inequities and the huge disparities in wealth an infrastructure that exist between the world

[14] Lee et al. p. 230

[15] Sunstein, Cass. *Republic.com*. Princeton University Press. New Jersey. 2001. p. 17

richest nations and the worlds poorest nations. Likewise there are inequalities other than those to which Sunstein refers; a disability such as blindness for example limits one's capacity to access the internet. Equally however the internet does address some other important inequalities such as language barriers. The benefits of the internet as a means of addressing the divide between the haves and have-nots is not as clear cut as Sunstein may have us believe and such inequalities do not relate solely to the internet but to technology in a broader sense.

As Johnson highlights, the digital divide is as much an issue of access to machines and software, as inequities in knowledge and expertise[16]. This is an important point particularly given the increasing role of computers in for example the education system. For a school to lack a resource such as computer laboratory is thus directly related to the educational opportunities that that school can in turn offer its pupils. It is not purely an issue of access to internet content that is at stake here. However with regard to information accessed via the web as an important good it is important to analyse some aspects of what makes information in general an important social resource. Johnson highlights that the concerns regarding inequities in access to the internet point to what exactly access to this good means, and this raises a few particularly important questions;

> What does access to the internet lead to? Why is access to computer and information technology so important? To what extent is this technology a means? If access is not equal or at least broad, exactly what kind of harm is done?[17]

I will not attempt to answer these questions however they point to an important aspect of ethical analysis of the internet i.e. what harms are posed by inequalities in access to the internet, or more broadly what harms are posed in a pandemic when inequalities in communication remain untouched. However more needs to be said about the nature of these inequalities.

2. Voluntary versus Involuntary Inequalities in Access

(a) Voluntary inequalities

Sunstein argues that the internet enables people to construct 'echo chambers of their own design' polarizing themselves towards certain pre-held views and creating the conditions whereby they can avoid opinions, views and interests which compete with their own. Sunstein's main argument is that new technologies make it easier for people to hear the opinions of like-minded but otherwise isolated others, and to isolate themselves from competing views[18]. For Sunstein a robust democracy requires that people interact with one another in a way that leads to debate and deliberation potentially with people who have quite different views, interests and claims.

Sunstein's account is worth bearing in mind when we consider the internet however it appears too that this is not some radical phenomenon unique to the internet. The concern is

[16] Johnson, Deborah. Computer Ethics (3rd edition) Prentice Hall. New Jersey. 2001. p. 219
[17] Johnson, 2001. p. 220
[18] Sunstein, p. 67

that this type of polarization and walling-off already occurs. Sunstein is correct that the internet is a new means of this happening but the issue is that this inequality is really nothing new. We may have people who simply avoid news media in television and print format the question is how can we supply these people with information when they appear to be refusing the resource? It is of course not clear cut what to do with such divisions it would certainly appear to be an impractical task to ensure that information reaches people who simply do not want it.

(b) Involuntary inequalities

As with more general barriers to information involuntary inequalities are those that arise not from an agent's choices or decisions but rather from unchosen circumstances. This second sub-set of inequalities in access to information appears to be more clear-cut and points to those inequalities that are not just rectifiable in a practical sense but ethically concerning in a way that voluntary inequalities are not. Involuntary inequalities can include disabilities, visual and hearing impairments and language barriers and can also include inequalities in socio-economic positioning. The value of the internet here appears to lie in its ability to provide information in a variety of formats which address involuntary hurdles to information.

Certainly some involuntary inequalities in communication can be addressed by the internet with positive ramifications for providing information in a pandemic. The question then is whether this particular virtue is sufficient to justify making the resource more equally available. Certainly some disabilities which the internet helps to relieve make it an important matter perhaps in the same way as a hearing aid helps a deaf person to better achieve their ends so too the internet might help to relieve some burdens presented by a diminished communicative capacity.

(c) The capacity of people to act on information

Another important point worth making here is that people vary in their capacities. It is one thing to have access to a resource and it is another to be able to make that resource into something of value. Beyond having access to information people need to be able to act on that information. In an emergency the issue is stark; providing messages and directives which tell people to act in certain ways must be tailored to an extent to be sensitive to what actions people are capable of. For example a message which tells people avoid work might be ignored by people who cannot afford to miss out on income. As Viswanath and Bond conclude in their analysis of communication and diet the capacity to act on information is of the upmost importance for promoting health and mobilising the public;

> Even if the messages are comprehended, the audience should have the material resources (such as being able to buy healthy foods, join a health club, live in a safe neighbourhood and have access to green space) to act on information.[19]

It is simply insufficient to disseminate generic information which is not sensitive to the capacities and realities of people's lives. It is one thing to tell someone an important message which can promote their welfare however it is another to take account of their capacity to act

---

[19] Viswanath, K and Bond, K. 'Social determinants and nutrition: Reflections on the role of communication'. *Journal of Nutrition Education and Behaviour*. Volume 39. Issue 2, Supplement 1. March-April 2007. p. 4

on that information. Telling women in Zimbabwe to practise safe sex by using condoms is for example an important health message however the concrete realities of many lives mean that some women don't adequately control this aspect of their autonomy. This can be seen as information having perhaps one function which requires action to support it for example supporting the safe sex message with increasingly tough punitive sanctions for those who violate another's sexual autonomy.

How then does this relate to the internet? On the one hand the internet can improve people's capacity to access information (i.e. addressing inequalities presented by disability or impairment), that is to say that information they would have otherwise missed out on can be made available with such technology. On the other hand that information will only have value if people can utilise it. As such the concern remains that the internet distributes some very poor quality information and as such does nothing to help people better achieve their ends. One issue here is that less well off communities for example often lack the public health infrastructure necessary to effectively combat a pandemic. It is one thing to recommend that people see a health professional it is another for health professionals to be easily accessible. In essence the claim here is that information is a necessary condition for meeting a pandemic however it is not sufficient and there are arguably other things that need to happen for information to be of real value to its recipients.

However the issue remains that communication is an important front when responding to a pandemic and a key concern is the harms that communicative inequalities open people up to. I have identified inequalities in access to information and sketched how the internet might play a role in addressing them however the jury remains out on whether the harms that the internet can mitigate require that the resource be made more accessible. I have raised the concern that communities without a sound communication infrastructure are vulnerable in a pandemic, as Garrets example highlighted. I have also outlined how the internet is both a potentially positive and a potentially negative force in pandemic communication. The central point I wanted to make was about the broader issue of inequalities in infrastructure to deal effectively with a pandemic and that lacking mechanisms for communication is symptomatic of a society or community which is severely disadvantaged.

5. Justice and Pandemic Communications

There appear to be no clear cut solutions to how to best utilise the internet my aim has been to draw attention to what inequalities in access to this technology actually mean in relation to a pandemic. Whilst the quality of internet information is an important concern the ethical dilemma cuts much deeper and relates to how societies lacking adequate communication are less well off and vulnerable in a pandemic. We ought to be concerned about the impact of a pandemic on groups who lack the important communicative means necessary to meet a pandemic effectively.

A fundamental issue in ethical debates surrounding pandemic influenza is the potential for existing inequalities to become exacerbated in an emergency. Lawrence Gostin raises the point that existing inequalities are likely to widen in the event of a public health emergency[20],

[20] Gostin, Lawrence O. 'Why Should We Care About Social Justice?' The Hastings Centre Report. July/ August 2007. 37, 4; Health Module. P. 3. p. 3

for instance it is likely that limits will be placed on people going to work, loss of income however is likely to impact more heavily on low income workers than it is more wealthy members of society. As such there is good reason to think that less well off groups may suffer more during a public health emergency with inequalities widening.[21]. Gostin draws attention to the Uscher-Pines study 'Planning for an Influenza Pandemic: Social Justice and Disadvantaged Groups[22] following on from the 2005 Bellagio Statement of Principles.[23] The main ethical concern raised by this literature is that groups and individuals in positions of disadvantage are often the most heavily affected during epidemics or pandemics. Uscher Pines et al's analysis of pandemic preparedness response plans argues that not enough is being done by officials to protect already disadvantaged groups, arguing that if countries do not incorporate the needs and rights of disadvantaged groups in planning, the effects of a pandemic may not only be catastrophic, but also catastrophically unjust[24]

If we know that the less well off are more likely to be further disadvantaged during an emergency involving infectious disease, this may create a moral imperative to ensure that less well off populations do not suffer the burdens in unfair or unjust ways, for instance incurring deaths that could have been prevented by more equitable arrangements.

Societies without sound or effective communication infrastructure will be vulnerable in a pandemic and important information may simply not get to the people who need it. We ought to be concerned about inequalities in communication when dealing with a pandemic and the initial justice concern here is that failure to address these inequalities will open people up to unjust harm.

The global nature of a pandemic and the potential impact on less well off societies who lack infrastructure is also a significant issue. The justice issue here becomes one of international aid requirements and the need for action at a variety of levels in order to protect less well-off and vulnerable communities. Whilst it is important that delivery of health care and other resources be an important component of this my argument is that communication is also a vital consideration. Ensuring that information reaches people will be an important matter of justice and helping vulnerable communities to develop pandemic responses and strategies for communication is imperative. A particularly important issue in pandemic ethics worth highlighting at the end of this discussion is the moral and ethical requirements which ought to be recognised at national and international levels to protect the less well-off. This discussion has highlighted that assisting such groups with their communication needs will be an important matter of justice. What we now face is the task of assigning and allocating responsibilities to achieve this end and strategies for assisting vulnerable population in the event of a pandemic.

CONCLUSION

I have tried to sketch out the ethical terrain surrounding the role of the internet, and communication in general, in a pandemic. In many ways this merely scratches the surface of a

[21] Gostin, 2007. p. 3

[22] Uscher-Pines L, Duggan PS, Garoon JA, Karron RA and Faden RR. *Planning for an influenza pandemic: Social Justice and Disadvantaged Groups.* Hastings Centre.

[23] *Bellagio Group: Statement of Principles.*

[24] Uscher-Pines, et al, p. 38

difficult and complex debate about the value of the internet however it also moves beyond this discussion to look at a specific situation where the internet is assumed to have potentially high value. This is by no means as exhaustive study of the issue(s) however I hope to have drawn attention to an issue that deserves more prominence in both the debate about justified responses to a pandemic emergency and the debate about the value of the internet namely the role of social justice.

I have tried to show what communicative inequalities are and how they can lead to particular harm(s) in a pandemic. The main point is that the digital divide is not just about the world-wide-web it is about inequalities in access to resources such as technology and know-how. It is about the infrastructure of certain communities and the need for adequate preparation to deal effectively with a pandemic emergency. These inequalities are most stark in less well off communities who may find themselves ill-equipped to deal with a pandemic if important resources are not available to them. These important resources include communicative apparatus such as media and public health officials disseminating information in very direct ways such as door-to-door. Information needs to get to people and the internet may be one avenue for doing this among many. We mustn't overstate its value but equally we must avoid downplaying the importance of communication.

REFERENCES

The Bellagio Meeting on Social Justice and Influenza. Bellagio Group: Statement of Principles. (2007) Available from:http://www.hopkinsmedicine.org/bioethics/bellagio (accessed 25.11.07)

Garret, Laurie. (2001) 'Understanding Media's Response to Epidemics', *Public Health Reports. 116(2).*

Gostin, Lawrence, O. (2007). Why should we care about social justice? *The Hastings Centre Report*; July/ August, *37*, 4, Health Module.

Hume, Janice. The "Forgotten" (1918). Influenza Epidemic and Press Portrayal of Public Anxiety. *Journalism and Mass Communication Quarterly;* Winter 2000, 77, 4.

Johnson, Deborah. (2000). *Democratic values and the internet.* From; Langford, Duncan (editor) *Internet Ethics.* MacMillian Press. *Hampshire and London., 187.*

Johnson, Deborah. (2001). Computer Ethics (3rd edition) Prentice Hall. New Jersey.

Lee, Connal & Rogers (2006). Wendy A. Ethics, Pandemic planning and Communications. *Monash Bioethics Review, Vol. 25,* No. 4, October.

Lee (2008). Connal Rogers Wendy A and Branauck-Mayer, Annette. Social Justice and Pandemic Influenza Planning: The Role of Communications Strategies. *Public Health Ethics.*Volume 1. Number, *3.* November 229-230.

Saliou, Pierre. (1994). Crisis communication in the event of flu pandemic. *European Journal of Epidemiology., 10*, 515- 517.

Selgelid, Michael. (2005). Ethics and Infectious Disease. *Bioethics., Volume 19*, Number, 3.

Smith, Charles B. Battin, Margaret, P. Jacobson, Jay, A. Francis, Leslie, P. Botikin, Jeffrey R. Asplund, Emily P. Domek, Gretchen J. & Hawkins, Beverly. (2004). Are there characteristics of infectious disease that raise special ethical issues? *Developing World Bioethics. Volume 4*, Number 1.

Sunstein, Cass. (2001). *Republic.com*. Princeton University Press. New Jersey.

Uscher-Pines, L., Duggan, P. S., Garoon, J. A., Karron, R. A. & Faden, R. R. (2007). *Planning for an influenza pandemic: Social Justice and Disadvantaged Groups*. Hastings Centre.

Viswanath, K. & Bond, K. (2007). 'Social determinants and nutrition: Reflections on the role of communication'. *Journal of Nutrition Education and Behaviour*. Volume *39*. Issue 2, Supplement 1. March-April.

In: Internet Issues: Blogging, the Digital Divide and Digital... ISBN: 978-1-60741-824-5
Editor: Cassie M. Evans, pp. 177-188

Chapter 11

UNDERSTANDING "SERIOUS VIDEOGAME" STORYLINE AND GENRE PREFERENCES RELATED TO GAME IMMERSION AMONG LOW-INCOME ETHNICALLY DIVERSE URBAN AND RURAL ADOLESCENTS

***Victoria Thompson*[1], *Debbe Thompson*[2] *and Tom Baranowski*[2]**
[1]The University of Texas, M.D. Anderson Cancer Center, Houston, Texas, USA
[2]USDA/ARS Children's Nutrition Research Center,
Baylor College of Medicine, Houston, TX, USA

ABSTRACT

Various media have been embraced by researchers and non-profit organizations because they help school-aged children acquire knowledge and develop decision-making skills. Serious videogames (i.e., games designed to entertain and educate, train or change behavior) may be effective at modifying youth health behavior and their related psychosocial influences. For serious video games to be effective, however, they need to be consistent with youth game-play preferences and expectations. It is important to understand player preferences so that immersive serious games can be developed that promote desired changes. While studies have reported ethnic and gender differences in time spent playing videogames and game-play preferences, few have examined regional differences in these variables. This chapter employed both quantitative and qualitative methods (i.e., surveys and focus groups) to examine preferences for storyline genres and plot content of non-violent videogames as well as computer access, knowledge and game-play frequency in a sample of predominantly low-income urban middle school students in Texas and rural middle school students in North Carolina. Survey results revealed that White and Other/Native American male students played videogames more frequently than African-American or Hispanic male students. Although focus group participants preferred action games with some violence, they had positive reactions to challenging adventure games with male and female characters of diverse ethnicities and narratives without overt sexual overtones. Incorporating storyline and game-play preferences may promote game immersion so that the player relates to and interprets the messages based on personal experiences. Critical thinking skills undergird and complement attempts to achieve positive behavior change.

INTRODUCTION

The popularity of videogames as measured by sales, prevalence, and usage has grown exponentially from the two-dimensional days of "Tennis for Two" and "Pong" to the cinema-like quality computer games of the 21st century. From 1996 to 2008, the unit sales growth of computer and video games increased more than 400% (from $2.6 billion to $11.7 billion) [1]. In 2009, approximately 70% of American households played video games [1]. The top five most popular game genres among teens were racing, puzzle, sports, action, and adventure, respectively [2]. Although children and adolescents have been reported to represent only 25% of players [1], video gaming was pervasive, with 99% of boys and 94% of girls [2] playing an average of approximately 50 minutes per day [3]. Approximately one-third of teens played at least once a day, while one in five played three to five days per week [2].

The technology on which to play video games was readily available to children and adolescents. Console devices such as Xbox and PlayStation were the most common modes of play, followed by personal computers, portable gaming devices and cell phones, respectively [2]. In 2004, most homes with 8 – 18 year olds had a computer and video game console [3]. More than half of children and adolescents owned a handheld video game player [3]. On average, children and adolescents spent 49 minutes per day playing console and/or handheld video games alone [3]. The "daily gamer" was a 12-14 year old male who played games online [2]. No regional differences have been reported in these practices.

Incorporating game elements that resonate with a target audience makes videogames more likely to appeal to a specific population. Transportation Theory posits that one is immersed into a story at the diegetic level (i.e., the game's story world) through mental imagery, emotional investment, and disregard for what is going on around them in the real world and may affect beliefs or attitudes within a narrative [4, 5]. Because games are not just narratives, transportation in a video game world includes the nondiegetic level (i.e., player's love of the game and the strategy with which they play it) of participatory activity in a virtual reality. Thus, the conventions of the game's environment must closely match player's expectations; the player's actions must have a significant impact on the environment; and the established conventions of the environment must remain consistent throughout the game. Therefore, serious videogames that have youth appeal, are perceived as personally relevant, and meet player game-play expectations are more likely to create an immersive experience. However, little is known regarding youth preferences for storylines or characters.

Youth reportedly play a variety of video games with age, gender, ethnic, and socioeconomic differences being reported in genre preferences. Twelve to 14 year olds preferred adventure and sports games more than 15-17 year olds [2]. Boys were more likely to play games from a wider range of genres than girls. Girls most preferred puzzle games while boys preferred action, sports and adventure games [2]. Similarly, African-American teens played sports and adventure games more than White teens [2]. African-American and Hispanic teens were more likely than their White counterparts to play fighting and survival horror games [2]. Lower-income teens were more likely to play racing, adventure and survival horror games than those from families making more than $50,000 per year [2]. White teens were more likely to play massive multi-player online games (MMOG) [2]. Ethnic and gender differences are well-researched; however, little is known about whether regional differences in types of games played exist.

This chapter examined computer access and knowledge, game-play frequency, preferred storyline genres and plot content of serious videogames as reported by predominantly low-income, urban, ethnically-diverse, middle-school students in Texas and rural middle school students in North Carolina to better understand the regional differences in these areas of interest.

METHODS

This study informed the design of two serious health-promoting videogames for 7th and 8th grade students [6]. Seventh and eighth grade urban and rural students were the target populations for this chapter which was a component of a larger multi-site primary prevention trial designed to promote healthy eating and physical activity behaviors in these age groups [7]. Two data collections were conducted at two sites (one urban, one rural) with non-remedial 7th and 8th grade students: a brief survey to assess familiarity and accessibility to videogames, and focus group discussions regarding narrative preferences.

Schools were selected based on both location and percent of enrolled ethnic minority students (>50%) in schools participating in a larger intervention study [7]. One inner city middle school in Texas agreed to participate (93 percent free/reduced lunch; 54 percent African-American, 42 percent Hispanic) as did one rural middle school in North Carolina (57 percent free/reduced lunch; 49 percent African-American, 49 percent White). The project was approved by the Institutional Review Boards of the Baylor College of Medicine and the University of North Carolina School of Nursing in 2004. Permission to conduct the research was received from the principal of each participating school prior to recruitment. Only students returning signed parental consent and personal assent forms participated in the survey and/or focus group discussions.

Survey

Video and Computer Gaming Practices and Preferences Instrument. The 16-item survey was adapted from previously published computer questionnaires and surveys [8, 9] by adding a checklist of popular computer- and video-games compiled by searching video- and computer-game websites. Most popular games were determined using website ratings for the most sold and most played videogames in 2003. The survey was evaluated extensively by project staff prior to use in the study. To reduce error from entering hand-written data [10], surveys were uploaded to personal digital assistants (PDA) (Palm, Inc. Santa Clara, CA) and completed anonymously in the same week as the focus groups. Completed surveys were downloaded into SPSS 11.0 (Chicago, IL) for analyses. Students could complete the survey and/or participate in the focus groups.

Survey Analysis. Descriptive statistics were presented by site (i.e., North Carolina, Texas), gender and ethnicity. The chi-square statistic was used to test for differences in responses by demographics. All analyses were performed in SPSS 11.0. 5

Focus Groups

Focus group discussions were conducted in the same schools as the survey. Depending upon the number of students needed, those agreeing to participate in both the survey and focus groups were randomly selected to participate in focus groups.

Focus Group Participants. All 7th and 8th grade students attending the participating middle schools were eligible to participate. Focus group participants received a small financial incentive ($20) in the form of cash or a discount department store gift card.

Description of Focus Groups. Focus groups were conducted at school, during the school day. All focus groups were conducted within one school week of each other. Participants were divided into groups on the basis of grade. An average of nine (9) participants was in each group (range 8 – 10). Focus groups were audiotaped in triplicate. The first author, a trained focus group moderator, led all focus groups. An assistant moderator was present to take manual notes, observe non-verbal cues and probe when invited by the moderator. Groups followed a structured script and lasted approximately one hour. The script was divided into sections: setting, characters, plot and game-play (Table 1). Probes and prompts were utilized to enrich, clarify, and expand student responses and to explore discrepant information across focus groups.

Table 1. Key Focus Group Questions and Areas of Interest

Areas of Interest	Focus Group Questions
SETTING	1. What are the first words that come to your mind after hearing this story? 2. Now that you've heard this story, describe where you think this story takes place.
CHARACTERS	3. Tell me about the characters.
PLOT	4. What do you think happens as the story continues. (After rating the story) Tell me your reasons for giving this rating.
VIDEO GAME PLAY	5. If this were a video game, describe the kind of video game you think this would be like.

Three storylines of different genres were discussed in detail in the four 7th grade and four 8th grade focus groups. One mystery and one adventure storyline audio recording was played to respondents in both grades. An additional comedy storyline was played to 7th graders. Instead of the comedy storyline, 8th graders heard a "drama" storyline. The moderator first read a short two-sentence description of each story followed by an audiotaped three-minute dramatized summary of the storyline. The discussion of the respondents' perceptions of the story and game components followed. Semistructured questioning following a script was used to ensure consistency in questions asked across groups. No more than two focus groups were conducted in one day. Audio storyline trailers, focus group questions and focus group structure were pre-tested with a small group of adolescent gamers (both male and female) prior to the focus groups. Scripted questions were revised based upon this discussion.

Focus Group Analyses. Within one hour of completing the second focus group of the day, the moderator and assistant moderator debriefed by discussing salient findings from each group [11]. The purpose of the debriefing was to capture the first impressions and main points of the relevant session, delineate key findings, and highlight and contrast findings with those from earlier focus groups. Responses from the focus groups were categorized by each storyline according to the previously mentioned foci of the discussion: setting, characters, plot, and game-play. During the debriefing session, the moderator generated a comprehensive key point summary (KPS) using these four (4) categorizes as an organizing structure. The emerging themes were the crux of the analyses. To ensure accuracy, a third party, who did not attend the focus group discussions or the debriefing, audited the KPS by listening to each focus group audiotape in its entirety and comparing the KPS to the discussion. No major changes occurred from this third party review.

RESULTS

Survey

A total of 98 Texas respondents (13.6 + .91 years; 44 percent male; 40 percent 7th graders; 58 percent African-American, 35 percent Hispanic, six percent Other/Native American, and one percent White) and 73 North Carolina respondents (12.5 + .65 years; 47 percent male; 80 percent 7th graders; 44 percent African-American, 40 percent White, 12 percent Other/Native American, and four percent Hispanic) completed the survey.

There was no difference in the presence of at least one home computer between Texas and North Carolina students. Among those reporting home computers, North Carolina students reported a somewhat greater percentage of home computers than did the Texas students (X^2=11.2, $p<0.05$). All Texas respondents and 97 percent of North Carolina respondents who had a computer at home reported having access to their home computer (Table 2). .

Favorite games were similar across sites. Students reported using numerous navigation tools: mouse; numbers, arrows, and letters on the keypad and joysticks. Joysticks were the least used navigation tool and more students reported using the mouse and arrow keys than other tools. There were, however, no regional differences in preferred tools. Across both sites (X^2 = 10.98, $p<0.05$), all White students (n=29) and Native American/Other students (n=14) reported having a home computer. Seventy-nine percent of African-American (n=66) and 78 percent of Hispanic (n=29) students had home computers. At 31 percent and 21 percent, respectively, White and Other/Native American respondents reported the highest number (i.e., three or more) of home computers ($X^2 = 19.14$, $p<0.05$).

No ethnic or regional differences emerged regarding self-reported knowledge of using a computer. Among males, more White (35 percent) and Other/Native American (60 percent) respondents played computer/videogames several times a day than African-American (8 percent) and Hispanic male respondents (7 percent) ($X^2= 35.64$, $p<0.05$). A significant association between when playing computer/videogames and race/ethnicity was also observed for males only ($X^2=17.63$, $p=0.040$). Only 7 percent of Hispanic males reported playing computer/videogames mostly on the weekends compared to 38 percent, 40 percent and 18 percent of African-American, Other/Native-American and White males, respectively. Twenty-nine percent and 22 percent of Hispanic and African-American male students reported playing computer/videogames after school compared to 6 percent of White and no Other/Native American students, respectively.

Focus Groups

Participants. A total of 76 students participated in 16 focus groups (i.e., eight groups per site; 59 percent African-American, 22 percent Hispanic, 14 percent White; 4 percent Other; 43 percent male) with some focus group respondents also completing the survey. In Texas, a total of forty 7th (n=20) and 8th (n=20) grade students (67 percent African-American, 30 percent Hispanic, 3 percent White; 45 percent male) participated in focus group discussions. Thirty-six 7th (n=17) and 8th (n=19) graders (50 percent African-American, 28 percent White; 14 percent Hispanic; 38 percent male) participated in North Carolina.

Setting and Characters. Action games were the student preferred game genre, regardless of grade, ethnicity, gender, or region. Although most responses regarding story settings and characters were story specific, students preferred an ethnically-diverse character cast comprised of both males and females. While characters were not required to be beautiful, attractiveness of the main characters appeared to be an important criterion. As one student reported: *"I ain't gonna...play [a video] game if [she's] ugly."*

The perceived ethnicity and physical traits of the characters were dictated by the voices of the actors and actresses performing the voice-overs (i.e., students interpreted the African-American-sounding voices as African-American characters, non-African-American sounding voices were non African-American characters, Latino accents were Hispanic characters, etc.). Voice was important to the participants, as captured in the following statement: [I decide what they look like] *"...based upon what they sound like..."* because if they looked otherwise *"[the story] wouldn't be believable.*

Students appeared to prefer characters that fit into traditional gender roles. For example, one participant stated: *"...one of the girls fall[s] and... the guy which is all muscular and stuff, he picks her up, throw[s] her over his shoulder and he start[s] to run."* Interestingly, however, male and female students at both sites supported the idea of a teenage female scientist as the lead role in the 8th grade storyline and stated that a teenage male scientist would *"...mess it up"*.

Table 2. Computer Survey

	NC (n = 73) n (percent)	TX (n = 98) n (percent)
COMPUTER ACCESS AND KNOWLEDGE		
Do you have a computer at home? (% yes)	62 (90%)	76 (80%)
Of those reporting the presence of a home computer		
How many computers do you have in your home?		
1	32 (51.6)	57 (75.0)
2	18 (29.0)	13 (17.1)
3	9 (14.5)	2 (2.6)
more than 3	3 (4.8)	4 4 (5.3)
Can you use a computer in your home? (% yes)	60 (96.8)	76 (100.0)
Do you have your own computer? (% yes)	27 (43.5)	33 (43.4)
How much do you know about using your computer?		
1=none	1 (1.4)	0 (0.0)
2=a little	7 (10.1)	11 (11.6)
3=some	33 (47.8)	44 (46.3)
4=quite a lot	27 (40.6)	40 (42.1)
GAMEPLAY FREQUENCY AND PLATFORMS		
How often do you play computer or video games?		
1 = several times/day	10 (14.5)	14 (14.7)
2 = everyday	15 (21.7)	21 (22.1)
3 = several times /week	14 (20.3)	23 (24.2)
4 = about once/week	16 (23.2)	11 (11.6)
5 = several times/month	6 (8.7)	12 (12.6)
6 = about once/month	5 (7.2)	11 (11.6)
7 = I don't play computer or video games	3 (4.3)	3 (3.2)
When do you usually play computer or video games?	weekends: 22 (33.3) after school: 9 (13.6) in the evenings: 7 (10.6) All of the above: 28 (42.4)	weekends: 25 (27.2) after school: 22 (23.9) in the evenings: 16 (17.4) All of the above: 29 (31.5)
When I play computer or video games, I usually play on…		
PC	16 (24.2)	29 (32.6)
MAC	0 (0.0)	9 (10.1)
Handheld Gameboy	0 (0.0)8 (12.1)34	4 (4.5)
Nintendo Gamecube	(51.5)	3 (3.4)
PlayStation 2	5 (7.6)	22 (24.7)
Xbox	3 (4.5)	18 (20.2)
Nintendo Entertainment Center/Super Nintendo		4 (4.5)

Plot. Although the games to be developed were slated to help players adopt healthy lifestyles, when asked for suggestions of what to include in a video game promoting healthy diet and physical activity behaviors to others their age, the initial response of students was to create an action-packed plot, with the expectation that "*someone should die*". When asked for an alternative storyline that did not include death or violence, the following seven themes emerged regarding the plot of a non-violent videogame that would appeal to teenagers: (1) a central character is captured and rescued; (2) the antagonist has an army; (3) a subtle or implied romantic component between a male and female character exists (an overt relationship was not recommended by most); (4) obstacles impede progress; (5) a central character acts as a traitor or double-agent; (6) a central character has an interesting history or genealogy; and (7) the story of the non-violent game should be centered around exploration and clue finding [6].

Gameplay. The most often repeated game-play component was "*levels*" of play. And, in the absence of violence, students recommended "*intellectual challenges*". For example, they recommended a game where players engaged in an interesting "*scavenger hunt*" with the goal of finding clues to solve a puzzle or overcome barriers.

CONCLUSION

Despite substantial concern about the inappropriateness of interactive multimedia games for limited-income populations because of the lack of availability of computers among limited-income children [12], these findings in both inner city and rural populations with substantially limited incomes demonstrate and are consistent with previous findings of the wide availability of home computers [3] and their accessibility to students, regardless of region. Virtually all participants in these studies played videogames frequently, with White males playing more frequently than African-American and Hispanic male students. This finding, in contrast to the findings of the Kaiser Family Foundation [3] where African-Americans and Hispanic students spent more time playing videogames than their White counterparts, may be partly attributed to our study of daily playing frequency versus minutes of play and gender extraction. Based upon our findings, White males play computer games on more days than African-American and Hispanic male students regardless of region. The seeming discrepancy between these two studies may be reconciled by perhaps Hispanic and African-American students playing less frequently than White middle school students, and when playing Hispanic and African-American middle students play approximately 10 to 20 minutes longer than Whites, respectively. Finally, our population was mostly female and males played approximately one hour longer than females [3].

Identifying genres that children and adolescents prefer is complicated. There is little consistency in categorizations among the industry and academics. For instance, while the NPD Report [13] indicated "Family Entertainment" was the favored genre among all games sold, the Pew Report did not categorize games into such a genre [2]. In our study, we created an action/ adventure game category, unlike these more widely recognized national surveys. Although our focus group participants preferred action games, they accepted challenging adventure games with male and female characters of diverse ethnicities and narratives without overt sexual overtones. Thus, these focus group results provided a first insight into child preferences for setting, character, plot and game-play of health promoting non-violent video games. Identifying and incorporating elements that resonate with the target audience should promote game immersion, thus enhancing the likelihood the audience will relate to, interpret and adopt the messages based on personal experiences.

While these samples were selected because they represented the target populations for a larger intervention study, and thereby were not nationally representative, the findings suggest that properly designed videogames could reach and would be played by ethnic minority children in both urban and rural areas. The high levels of videogame use among these children [3] suggest that videogames may be an important medium for providing health behavior-change experiences to children. Since videogames are a sedentary activity, concern exists regarding their use to promote physical activity. Alternatively, if students play games anyway, spending some of that time to promote physical activity and better dietary habits, as suggested by the Institute of Medicine [14], may be beneficial. Further, The Kaiser Family Foundation study suggested that care be taken when concluding that "time spent with media is synonymous with time taken from other activities" [3] as this is not always the case [15]. However, using games for edutainment purposes may be a viable approach to decrease the displacement of academic pursuits that games are known to have [15], particularly in students who play more during the week than weekends. The advent of the "serious videogame" genre has provided effective strategies for promoting healthy behaviors such as fruit and vegetable consumption among children [16], diabetes management among children and adolescents [17] and gait improvements among multiple sclerosis patients [18]. Designing games to increase physical activity should mitigate these concerns.

This chapter examined student preferences and access in both urban and rural schools; nevertheless, lack of generalizability to the general population still exists. Although we measured the presence of home computers and inferences can be made from the responses regarding preferred videogames and platforms, we did not explicitly survey the presence or ownership of videogames, or videogame platforms or home Internet access. At the time of data collection, game consoles such as the Wii by Nintendo were not available and cell phones were not as ubiquitous among adolescents [19]. Consequently, preferences and practices among currently popular game consoles and cell phone games, known to be more prevalent among girls, black and lower-income teens [2], were not explored. Further, the joystick being identified as the least preferred game navigation tool (i.e., mouse, keyboard, etc.) is likely a fallacy as preferred tools are likely dictated by the requirements of the most popular games. Again, comparing preferred genres to those found in other surveys is complicated because of the inconsistencies in categorization.

We found no other research that examined the preferences for computer- and videogame narrative and characters of a non-violent videogame among middle school students. The storyline selected by students was action-adventure over the comedy, drama and mystery options at both sites, both genders, and all ethnicities. This reflects a high level of preference for action-adventure games [20, 21]. Child expectations for action in the storylines reflected the prevalence of action in cinematic movies [22].

Interestingly, student character preferences did not totally reflect characters of popular games as students preferred an ethnically-diverse cast of characters. One study examining over 1,700 characters in the 10 top-selling video games for seven different game systems (i.e., Game Boy Advance, Game Boy Color, Dreamcast, Nintendo 64, Playstation, Playstation 2, and the personal computer) found little to no ethnic or racial diversity and games for older children reinforced stereotypes [23]. Games designed for young children only featured White characters (23). Of all games, approximately 60% and 90% of the human characters and heroes, respectively, were White [23]. White female characters outnumbered female characters of every other racial group in games. No game included Latina characters. Increasing character diversity, particularly in serious videogames, may minimize the negative social messages these images propagate in mainstream media as well as promote game immersion through character identification.

Games promote gender stereotypes [23] and, to some degree, this population also preferred these characterizations such as "muscular" male rescuers and female characters requiring rescue. However, because students specifically preferred no overt sexual storyline and an intelligent female lead character, it appears stereotypes that present characters in a negative light did not appeal to the students participating in this research. For instance, students did not mention as a preference accentuating female sexuality and female characters who are verbally aggressive (i.e., verbal insults, ridicule, screaming), both of which are stereotypes propagated by games (23). Students disavowed overt sexual content, instead preferring the subtle innocence of teen love interests.

Designing immersive and engaging "serious games" that eliminate or minimize violence is a challenge for game developers, particularly when violence in videogames played by youth continues to receive such negative reactions. Games in which the player is challenged to find clues to overcome barriers and proceed to the next level is a common mechanic in videogame play and, in the absence of violence, was recommended by focus group participants. Using these principles to design serious video games would likely appeal to children and thereby overcome key barriers to attentiveness to health behavior-change messages.

Computer-based interactive media appear to be appropriate for reaching large numbers of middle school children, even ethnic minority children, in both rural and inner city areas. The action-adventure story appears most likely to meet their interests and thereby provides an important vehicle within which to design health behavior-change games. An immersive non-violent adventure game in which the player proceeds to different levels should also include a central character who is captured and rescued amid an antagonist with an army and numerous obstacles. Properly designed serious games hold promise for enhancing education and health promotion efforts.

REFERENCES

[1] Entertainment Software Association. (2009). 2009 Sales, Demographic and Usage Data: Essential facts about the computer and video game industry.

[2] Lenhart, A., Kahne, J., Middaugh, E., Macgill, A. R., Evans, C., & Vitak, J. (Sept. 2008). Teens, video games, and civics: Teens' gaming experiences are diverse and include significant social interaction and civic engagement. Pew Internet & American Life Project. Washington, DC.

[3] Rideout, V., Roberts, D.F., & Foehr, U.G. (2005). Generation M: Media in the lives of 8-18 year-olds. The Henry J. Kaiser Family Foundation. Menlo Park, CA.

[4] Green, M.C., & Brock, T.C. (2000). The role of transportation in the persuasiveness of public narratives. *Journal of Personality and Social Psychology, 79,* 701-721.

[5] Green, M.C., & Brock, T.C. (2002). In the mind's eye: Transportation-imagery model of narrative persuasion. In M. C. Green, J. J. Strange, & T. C. Brock (Eds.), *Narrative impact: Social and cognitive foundations* (pp. 315-341). Mahwah, NJ: Erlbaum.

[6] Thompson D, Baranowski T, Buday R, Baranowski J, Thompson V, Jago R, and Griffith MJ. Serious Video Games for Health: How Behavioral Science Guided the Design of a Game on Diabetes and Obesity. *Simulation & Gaming, Dec 2008;* doi:10.1177/1046878108328087

[7] Venditti, E.M., Elliot, D.L., Faith, M.S., Firrell, L.S., Giles, C.M., Goldberg, L., et al. (2009). Rationale, design and methods of the HEALTHY study behavior intervention component. *International Journal of Obesity(Lond), 33*, S44-51.

[8] Bleakley, A., Merzel, C.R., VanDevanter, N.L., & Messeri, P. (2004). Computer access and Internet use among urban youths. *American Journal of Public Health, 94*(5):744-746.

[9] Fromme, J. (May 2003). Computer games as a part of children's culture. *International Journal of Computer Game Research*, *3*(1). Retrieved from http://www.gamestudies.org/0301/fromme/.

[10] Shaw, R., Coia, J.E., & Michie, J. (1999). Use of bar code readers and programmable keypads to improve the speed and accuracy of manual data entry in the clinical microbiology laboratory: experience of two laboratories. *J Clin Pathol.* 52(1):54-60.

[11] Krueger, R. (1988). Focus Groups: A practical guide for applied research. 1st ed. Newbury Park, CA: SAGE Publications, Inc.

[12] Becker, H.J. (2000). Who's wired and who's not: children's access to and use of computer technology. *Future Child.* 10(2):44-75.

[13] Entertainment Software Association. (2009). Sales and genre data. Retrieved from http://www.theesa.com/facts/salesandgenre.asp.

[14] McGinnis, J.M., Gootman, J.A., Kraak VI, eds. (2006). Food marketing to children and youth: threat or opportunity. Washington, D. C.: National Academies Press.

[15] Cummings, H. M. & Vandewater, E. A. (2007). Relation of adolescent video game play to time spent in other activities. *Arch Pediatr Adolesc Med, 161*(7): 684-689.

[16] Baranowski, T., Baranowski, J., Cullen, K.W., Marsh, T., Islam, N., Zakeri, I., et al. (2003). Squire's Quest! Dietary outcome evaluation of a multimedia game. *American Journal of Preventive Medicine,* 24(1):52-61.

[17] Brown, S.J., Lieberman, D.A., Gemeny, B.A., Fan, Y.C., Wilson, D.M., Pasta, D.J. (1997). Educational videogame for juvenile diabetes: Results of a controlled trial. *Medical Informatics (London).* 22(1):77-89.

[18] Baram, Y., & Miller, A. (2006). Virtual reality cues for improvement of gait in patients with multiple sclerosis. *Neurology.* 66(2):178-181.

[19] Lenhart, A., Arafeh, S., Smith, A., & Macgill, A. (2008) Writing, technology, and teens. Pew Internet & American Life Project. Washington, D.C.

[20] Funk, J.B. (1993). Reevaluating the impact of videogames. *Clin Pediatr (Phila).* 32(2):86-90.

[21] Haninger, K., Ryan, M.S., & Thompson, K.M. (2004). Violence in teen-rated videogames. *MedGenMed.* 6(1):1-11.

[22] Thompson, K.M. & Yokota, F. (2004). Violence, sex and profanity in films: Correlation of movie ratings with content. *MedGenMed.* 6(3):3-13.

[23] Children Now. (Dec. 2001). Fair Play? Violence, Gender and Race in Video Games. Retrieved from http://publications.childrennow.org/publications/media/fairplay_2001b.cfm

In: Internet Issues: Blogging, the Digital Divide and Digital... ISBN: 978-1-60741-824-5
Editor: Cassie M. Evans, pp. 189-197

Chapter 12

INCREASING UNSTRUCTURED INFORMATION REUSE

Rogerio Atem de Carvalho, Ronaldo Amaral Santos and Vanderson Mota dos Santos*

Information Systems Research Group (NSI), Instituto Federal Fluminense, Campos, Brazil

ABSTRACT

The increasing volume of electronic information available in searchable repositories has been facilitating the construction of new knowledge through information reuse and derivation. However, when it comes to unstructured information, such as text documents, retrieving whole documents, searching them "by hand", and copy-and-pasting the interesting parts is the most common form of information reuse. When it comes to more sophisticate objects, such as video and animations, the user performs a metadata search, retrieving whole potential objects. It is only by playing them that the reusable parts can be identified. The aim of this chapter is to show how it is possible - and necessary - to improve productivity to the task of unstructured information reuse by serving smaller pieces of documents and media, and also allowing users to build and share new objects from these pieces through the Web. Aiming to better present this idea, a tool designed to extract elements such as paragraphs, tables, images and references from documents; images from videos, and graphical objects from animations, is presented. This tool also supplies automatic document construction, allowing users to immediately create and share editable documents made from other documents' pieces.

1. INTRODUCTION

The increasing volume of electronic information available in searchable repositories has been facilitating the construction of new knowledge based on information reuse and derivation. However, when it comes to unstructured information, such as text documents,

* Corresponding author: ratem@iff.edu.br, rsantos@ iff.edu.br, vanderson.mota@gmail.com

retrieving whole documents, searching them by hand, and copy-and-pasting the interesting parts is the most common form of information reuse. This means that one is supposed to retrieve and check whole documents, when often it is only parts of them are really interesting. When it comes to more sophisticate objects, like video and animations, the situation is even worse: the user must search based solely on metadata, retrieve the whole objects – sometimes quite big – and only by playing them is it possible to identify reusable parts. This level of effort is opposite to the desire of the current demographics of the Internet: "a generation of professionals who want bite-sized chunks of information that will enable them to get on with whatever it is they have to do" [1].

Seeking to provide better reuse of these smaller chunks of information, a tool named Granulator is here presented. This tool is part of NSI[2], an Enterprise Content Management solution based on open source software [2], and aims to extract elements such as tables and images from documents, images from videos, and objects from vector graphics. The user searches text-based content using keywords, and searches videos and graphics using image similarity. After searching, the user can select "grains" – as they are called in this context – and put them into a shopping cart ("Grain Cart"). From the grains in the cart, an editable document is built, automatically referencing the grains' source documents (or media), allowing users to immediately create and share new content built on the reusable parts of other documents.

Throughout this chapter, a "document" refers to any unstructured information container, such as office documents (text, spreadsheet or presentation – all of these referred to as "text"), video or animation. The term "grain" will refer to every reusable part of a document that is somehow delimited, such as paragraphs, tables, images, formulas and references in texts; individual frames in videos; and individual objects in Scalable Vector Graphics (SVG). This document is organized as follows: after this introduction, grain extraction from different document formats is explained; followed by explanations on how grains can be stored and served; and finally some conclusions.

2. Extracting Document Grains

The goal of separating content in smaller parts is to offer a better user experience, by allowing the composition of new content based on the reuse of parts of objects stored in content base. Therefore, a grain is defined here as a part of a document that can be extracted and served individually, improving reuse and reducing the resources necessary for serving content. These resources include network resources and also time resources, since the user doesn't have to open each document and search for prospective reusable parts.

Granulator is implemented in the Python language and wrapped as an XML-RPC callable service. It can receive different types of content streams such as Mark Up, namely XML (including Scalable Vector Graphics - SVG), HTML, and compressed XML (Open Document Format – ODF); textual binary (Microsoft Office and PDF); and video (AVI, WMV). These streams are parsed for grain searching in different ways, implemented using the Strategy design pattern to facilitate future inclusions of new types of documents or grains. Although Granulator always returns XML grains, with images wrapped with their metadata, each type of content presents particularities, which are presented in the next sections.

2.1. Parsing Text Files

Parsing text documents starts by using an appropriate converter. In the case of MS-Office or Open Office Org (O3), an O3 server was implemented in the form of a service able to perform every operation performed by a local O3 instance, including converting MS-Office documents from their binary format into ODF. This service is implemented in a very scalable way, capable of handling hundreds of simultaneous requests.

In the case of PDF documents, the free PdfToHTML converter (http://pdftohtml.source forge.net/) is used, however, in a near future, the O3 server will be employed, since O3 community is implementing PDF converting facilities into the suite. When Granulator gets a text document, it identifies its type and sends to the appropriate converter, getting a XML representation of the document.

The next step is to parse the document's XML structure for identifiable grains of information. These grains inherit the metadata from the owner document and also have their own metadata (such as format, size etc). Any text in the grain is automatically indexed, meaning that it can be searched separately from its document. For instance, an image with a specific caption inside a text document can be found through its owner document's full text and metadata, but also by its own metadata and caption text. When a user opts for a granular search, he or she will get a results page with a hierarchy of documents and their respective grains.

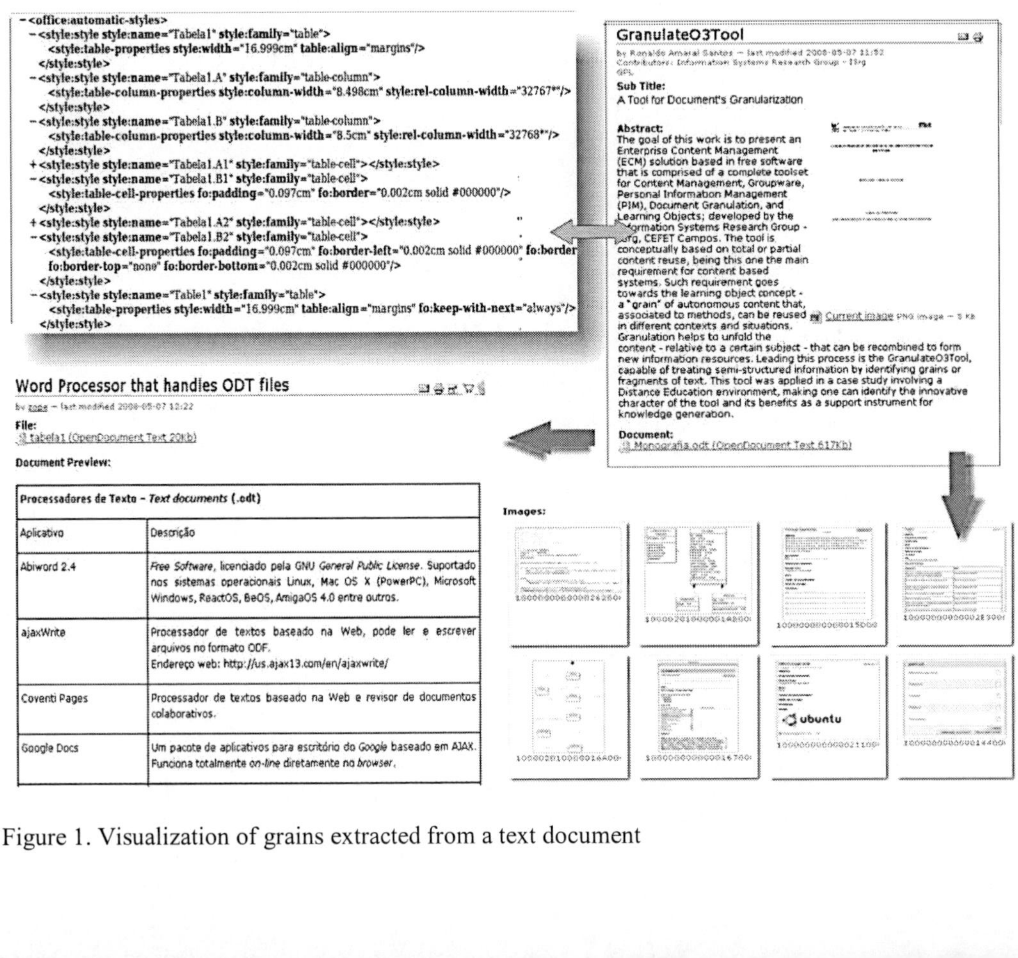

Figure 1. Visualization of grains extracted from a text document

2.1.1. Parsing PDF tables

Adobe Acrobat doesn't interpret tables as such, but as a set of text cells with boxes around. Therefore, when the package PdfToHTML generates a XML representation of a PDF file, the parsing algorithm infers which boxes are in fact table cells and builds a table by joining them. Although there are some commercial tools capable of doing PDF table extraction, Granulator follows the basic requirement of NSI[2] that all components of the system must be free/open source software. To this end, an improved version of the two-stage heuristic developed originally by [3] has been made which is capable of identifying most tables in a PDF file - although it is susceptible to specific PdfToHTML shortcomings. Hence, although other more robust heuristics were investigated, such as the one proposed by [4], to simplify the solution architecture and take advantage of community developments, the tendency is to use only the O3 in the medium term.

2.2. Parsing Vector Graphics

Granulating vector graphics is also done through XML parsing – in SVG format, with a difference: the parser looks for specific tags that identify the composite objects to avoid granulating very simple objects such as lines and backgrounds. Instead the parser searches for more complex objects, as shown in Figure 2, according to a heuristic developed specifically for the Granulator.

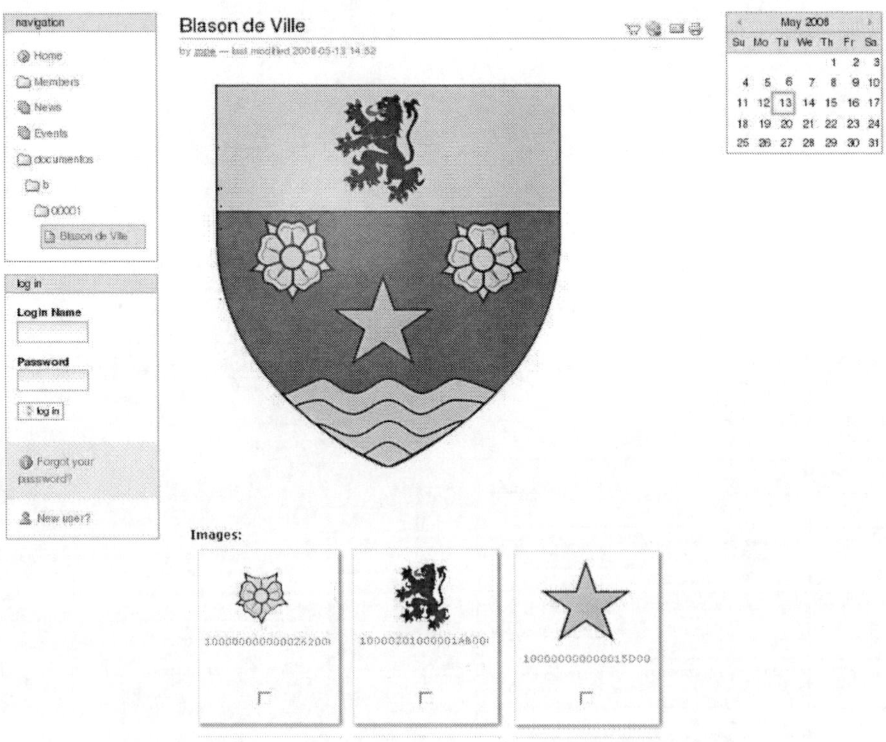

Figure 2. Visualization of grains extracted from SVG

Since SVG files are in essence tagged text, parsing these files also builds up text and tag indexes. In that way, besides image searching, an end user can search SVG files using their embedded text and functionalities. For instance, the user would be able to make queries such as *find an animation that has the text "education" and implements a Gaussian Blur effect*. However, since SVG also intensively uses path objects (set of interconnected points), an algorithm which uses an element's graphical bounding box is in development. This algorithm allows a user to interactively select parts of the graphical object and reuse them. This selection is done by defining a polygon by an arbitrary set of points marked using the mouse.

In the case of Flash animations, the content provider must convert them into SVG, since due to the lack of free and mature converter packages, this solution was not implemented into Granulator.

2.3. Parsing Video

Video granulation can be done automatically, by the detection of scene changes, using Data Compression and Movement Detection algorithms; or "manually", by extracting frames using a user-defined time interval. It is also possible for a content provider to define which frames best characterize the video and upload them together with the video as part of its metadata. Moreover, if the video is accompanied by a legend file, a fulltext search can be performed on the legends, allowing for the identification of the scenes related to the searched keywords.

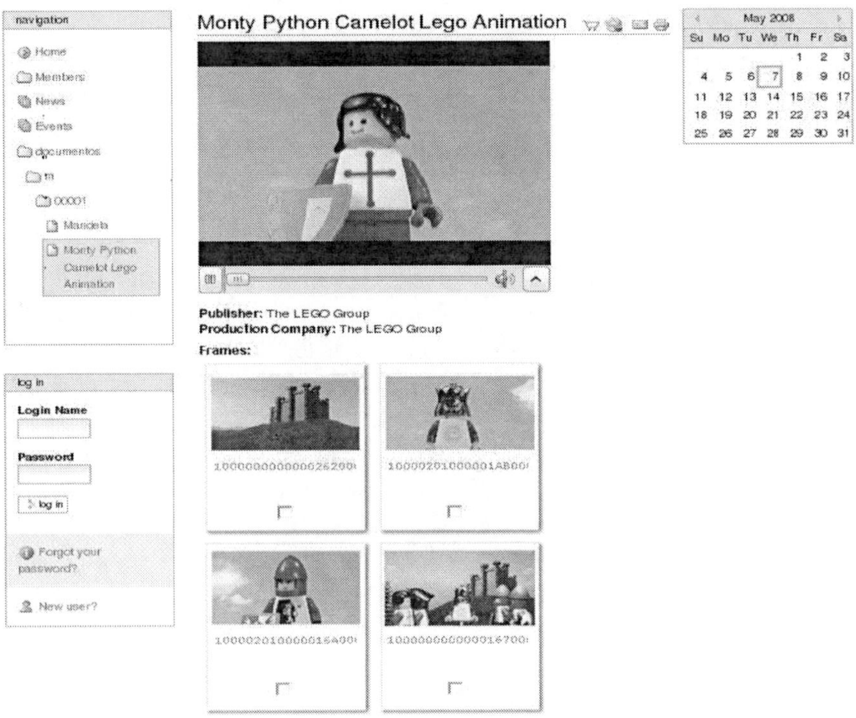

Figure 3. Visualization of grains extracted from a video

3. STORING, INDEXING AND SERVING

With the aim of exemplifying how grains can be treated after extracted, the next topics will briefly explain how NSI^2 stores and serves grains, using its presentation layer to direct requests to the different services presented hereinafter.

3.1. Storing Grains

NSI^2 implements a flexible storage service, which uses configurable criteria to allocate space for documents and grains in the file system. This service takes a document and returns an identifier (id) for it, used to access this document when needed. In that way, the complexity of storing the objects is hidden from the client application, which only needs to request a file by its id. The service keeps a File Allocation Table that maps ids to the files' and grains' physical locations, making the appropriate updates on this table when files are moved due to changes in the storage policy. Currently, file system based storage is used, but creating a relational database version is relatively easy.

In the case of grains, the storage service receives a XML stream that is a collection of grains, and stores them in a directory named after its owner document's id. For instance, if the document id is 12345, and its title begins with an "a", its grains are stored in a directory named /storage_root/A/12345/. Grains can be cached and served separately from their owner documents, reducing I/O and network traffic [5].

Users have the option of saving documents in HTML, ODF, PDF (generated by the O3 service), or in its original format. Grains can be saved in HTML or ODF. The system administrator is responsible of configuring these and other options, in accordance to the enterprise content reuse policy.

3.2. Indexing Grains

NSI^2 implements an indexing service, which takes as input the document and its grains' full text and metadata, when these are applicable. A grain's index entry also keeps a pointer to its owner document. Therefore it is possible to search for documents only, documents and grains, or only grains. In the latter case, the search is made on both document and grains metadata, but only grains are returned.

It is also possible to configure searches that work only on grains metadata. Granular search can be restricted by any metadata, such as format, type, date of upload or author, to reduce the result set size. Moreover, if a domain ontology is available, the user can opt for an enhanced granular search, making the grains inherit the semantic network associated to their owner documents.

Graphical grains, extracted from documents, vector graphics, animations and videos are searched using the Lucerne's search engine algorithms (http://lucene.apache.org/). The user uploads an image that is used for pattern matching, and the service returns a collection of images sorted by relevance. For the sake of improving the performance of this specific search engine, all images are converted into JPG format, but in the case of SVG files, they are served in their original format.

3.3. Serving Grains

NSI^2 offers a set of functionalities for reusing unstructured information. Self-Service was inspired by solutions like SafariU (http://www.safariu.com), which let users build new textbooks from parts of existing ones, providing automatic tables of contents and indexes [6]. However, it goes further by offering a fine-grained reuse mechanism, as well as more options on what to do with grains and whole documents.

Self-Service allows a user to search a NSI^2 catalog and tick the documents and grains of interest and put them in the grain cart, or in his or her personal file cabinet. In the case of grains, the user can select which types he or she wants to see: tables, images, paragraphs, formulae or bibliographical references. The grains and documents in the cart can be downloaded as separate files in a compressed file or can be used to create a new ODF or HTML document containing all objects, with a automatically generated reference list at the end of this document. A service, named DocBuilder, is used to build new documents on the fly and extract metadata for this reference list. After being generated, this document can be downloaded, edited with a built-in text editor, or placed in the user's file cabinet. With this document the user will have a basis for creating new content, like reports, articles, news, and learning material. The user also has the option of immediately sharing the generated file (compressed, HTML, or ODT) with other users, by allowing access to it from his or her cabinet or by submitting to be published in the main repository.

Since the quantity of grains returned can be overwhelming, the presentation tier of NSI^2 provides a publishing workflow that allows content publishers to delete non-reusable objects, such as bars, markers and such; and the grains are visualized as thumbnails that can be zoomed out, so that the final user doesn't need to download the originals to inspect their content (Figures 1, 2, and 3). Also, the rich set of metadata used for both documents and grains helps reducing the searching universe in advanced search options. Currently both Dublin Core (http://www.dcmi.org) and IEEE Learning Objects Metadata (http://projects.ischool.washington.edu/sasutton/IEEE1484.html) standards are pre-configured into NSI^2; however hybrid metadata sets can also be implemented, easing the integration of digital and learning objects libraries [7].

4. CONCLUSIONS

Granulator is a mature application, able to process batches of tens of thousands of documents by using commodity hardware in an effective way. In terms of use, Granulator is one of the services offered by a set of content management solutions based on NSI^2, which is in use by the Brazilian Ministry of Education's Secretary for Professional and Technological Education (Setec/MEC) in three integrated applications:

- Digital Library (articles and thesis repository): allow users to build new articles based on existing ones, and edit them on-line.
- Digital Document Center (administrative documents repository): allow users to build reports based on existing documents and templates.
- Learning Objects Catalog (repository for teaching material): new learning material can be built from mixing existing ones with new features.

Furthermore, a XML Warehouse [8] used to integrate all data managed by these applications will be built by the intensive use of granulating functions, currently in development. In that way, Granulator, together with other services, can provide a better experience for users, by offering ready-to-use pieces of semi-structured information supported by a rich metadata set, and allowing the construction of new content by mixing these pieces with brand new content.

Additionally, DocBuilder is evolving towards a more complete tool called the Digital Products Line (DigiProL), which will offer a complete set of through-the-web editing features for XML based document formats such as ODF and SVG. The main goal of this new tool is to allow a user with very few computing resources (a computer connected to the Internet and a browser), to reuse and create new knowledge. NSI^2 is also the basis of the digital library of the biggest Brazilian government digital inclusion program, which comprises of more than 700 community computing centers spreadout through the country, including some located at semi-isolated Amazon villages, which still have low connection to the Internet.

ACKNOWLEDGMENTS

The authors thank all NSI^2 collaborators for supporting this work and Setec/MEC for sponsoring it.

REFERENCES

[1] Reisman, S. (2007). Fast-Tracking Content: The Dramatic Evolution of Intellectual Property. *IEEE IT Professional*, *Vol. 9*, Num. 5, 2007. 58-60.

[2] De Carvalho, R. A. (2007). An Open Source Enterprise Content Management Solution Based on Open Source. *Research and Practical Issues of Enterprise Information Systems II*. New York : Springer-Verlag New York Inc, v. 1173-184.

[3] Yildiz, B. (2004). *Information Extraction – Utilizing Table Patterns*. Master Thesis. Technological University of Vienna.

[4] Liu, Y., Bai, K., Mitra, P. & Giles, C. L. (2007). TableSeer: Automatic Table Metadata Extraction and Searching in Digital Libraries. *In Proceedings of ACM IEEE Joint Conference on Digital Libraries*, (JCDL'07), Vancouver, Canada.

[5] Khaing, A. A. (2005). Ni Lar Thein. Efficiently Creating Dynamic Web Content: A Fragment Based Approach. Proceedings of the 6th Asia-Pacific Symposium on *Information and Telecommunication Technologies*.

[6] Buxton, S. (2007). Beyond search: content applications. *IEEE IT Professional*, *Vol. 9*, Num. 1, 29-35.

[7] Arapi, P. Moumoutzis, N. & Christodoulakis, S. (2006). ASIDE: An Architecture for Supporting Interoperability between Digital Libraries and eLearning Applications. Proceedings of the Sixth IEEE International *Conference on Advanced Learning Technologies*, 257-261.

[8] Nassis, V., Dillon, T. S., Rajugan, R. & Rahayu, W. (2006). An XML Document Warehouse Model. Proceedings of the 11th Int. Conf. on Database *Systems for Advanced Applications*.

In: Internet Issues: Blogging, the Digital Divide and Digital… ISBN: 978-1-60741-824-5
Editor: Cassie M. Evans, pp. 199-214

Chapter 13

TWITTER: A PRODUCTIVE LEARNING TOOL FOR THE SMS GENERATION

Adelina Moura** and Ana Amélia Carvalho
Institute of Education, University of Minho, Portugal

ABSTRACT

Twitter entered into the microblogging family and it is becoming a communication, information, and productivity tool. By enabling people to write messages up to just 140 characters it requires great information management and makes knowledge sharing fast and more efficient. Twitter as a learning tool can be used in different contexts. We can convert it into a tool for collaborative or individual learning. Making these assumptions we conducted an experiment based on microblogging with Twitter inside and outside the classroom, as a complement to Portuguese lessons. The activities were intended to help students to improve their writing skills, to develop reasoning, to improve their capacity for summarise, and to increase creativity. To achieve this students created micro stories and micro biographies, wrote Haikus, wrote a diary or poetry, published news, proverbs, thoughts, and aphorisms. These were some of the activities published by students on the classTwitter account.

Keywords: Twitter, SMS generation, learning tool, communication

1. INTRODUCTION

It is time to stop and reflect on how communication technologies have progressed in recent years. We have passed from e-mail to chat, blog, social networks and now to Twitter. Each new technology offers improvements for the user relative to its predecessors.

Jack Dorsey, the creator of Twitter defines it in a simple way: "Twitter is not a social network, but a new form of communication" (apud Frias, 2009), because Twitter follows the

* Corresponding author: adelina8@gmail.com, acc@ie.uminho.pt

model of taxi drivers' radio by communicating minute-to-minute. Inspired by this he transferred into the microblogging world the need for users to tell everyone what they are doing, or what they have in mind in real time.

Similarly for Barrett (2008), Twitter is "primarily a communication tool and has often been described as filling the gap between email and instant messaging (IM)". To Iskold (2007), Twitter is a "broadcast with an instant feedback", and he adds "that it is both a natural step from blogging and a weird experiment normally found in neuroscience Labs".

Despite its simplicity it has become very popular in a short period of time and its success is attributed, in its co-founder Biz Stone's (apud Brandon, 2009) words, "to the fact that people gravitate towards tools that can help them connect and share information. Beyond that, simplicity is probably the big factor in Twitter's increasing appeal". It is a kind of SMS-based Twitter. According to Martin (2009) its popularity is also due to the enthusiastic way that important public figures have taken to using this tool, "Twitter has grown exponentially in the past few months with no small thanks to celebrity exposure".

Twitter has grown over 1.382% in one year, according to Nielsen Wire (McGiboney, 2009). Despite this growth, Nielsen conducted a study on social networks' greater growth in recent months and found that 60% of Twitter users did not re-enter the system during the next thirty days. Only 40% of users remained activite on this tool.

Twitter has entered into various sectors of society. It is used commercially by companies like Dell, JetBlue and Whole Foods and is becoming the "center" of what happens in the world, says the co-founder Biz Stone in an interview with Computer World Blogs (apud Brandon, 2009).

Whereas blog posts are long, Twitter allows fragmented sentences, and can be published quickly and treated. It is this interaction between speed and quantity that has created a qualitatively different experience (Iskold, 2007). For this author, "Twitter pushed us all to the edge of real communication. Any more real would probably be telepathy!" People collaborate on Twitter in real time, to discover news, to find out what others are doing or even to ask or receive advice.

With regards to education, it is becoming a learning tool with multiple applications. With the reduction of text to a few characters, Twitter is now a valuable tool for learning, because it is necessary to reduce thoughts to the basics and creates a "mental gym", which makes it a cognitive tool, in Jonassen's opinion (2000). Twitters' economy of words has even more value in a world full of accessories, as mentioned by Querido (2009), a Portuguese journalist.

With the ability to add a mobile number to a Twitter account we can send and receive messages on the move. Mobile technologies allow a connection anytime, anywhere and are shaping the relation between users and the information that they can access on the move. Mobile phones have become indispensable objects, offering an endless variety of new and powerful features.

Among existing technologies, the mobile phone is the most widespread in the SMS generation which regards it as an extension to, or organic part of, the body (Oksman & Rautiainen, 2003). The spread of SMS writing plays a big role in Twitter's entry into mobile technologies. With Smartphones that enable web browsing, writing and reading, tweets has become more comfortable on smaller screens.

Twitter is one of the most appropriate tools for publishing micro contents because it has good convergence with mobile technologies. Its popularity led to the establishment of the

"Shorty Awards"[1]. This is a prize which aims to identify the best twitters on the planet, and it is ganning a lot of interest into the Twitter community. We believe that Twitter will emerge as a new educational tool in the future.

Taking all of the above into account, in this paper, we present details of a teaching experiment which used Twitter to complement the Portuguese lessons of 10th grade students. The activities were intended to assist and train students with their writing skills, to develop reasoning and analytical competence, and to improve their capacity for summarize and creativity, research activities, analysis and information sharing. Students had to write micro stories and biographies, create Haikus[2], write a diary, poems, and post news, thoughts, proverbs, aphorisms and riddles. These are some of the activities that were done by the students both inside and outside the classroom and published on Twitter.

2. Twitter: A Web 2.0 Tool for Learning

After the spread and success of social networks now is the time for microblogging applications to enter into education. To Zago (2008) microblogging is a kind of "simplified blog", because essentially it has the blog format, but in a simplified form. The main difference seems to be size limitation, because each update can only be up to 140 characters. Twitter is the most serious social web after Google, YouTube and Facebook and it is now the most popular microblogging tool on a global scale (Zago, 2008).

York (2007) presents ten ways to use Twitter:

- as a source of news and views
- a knowledge network
- "virtual water cooler" - a time of "chatter"
- keeping up with friends
- daily trips
- monitor events
- tool for marketing or public relations
- learning tool
- as entertainment
- as a daily lesson in humility and succinctness

The Centre for Learning & Performance Technologies[3], through its list of the 100 best learning tools in 2009, allows us to know the opinion and selection tools for education by professionals involved in teaching and training. Among all these tools, Twitter is one of them. Jane Hart (in Esains, 2009) suggests several reasons for the popularity of Web 2.0 tools for

[1] The Shorty Awards honour the world's top Twitters. We can nominate as many people for as many categories as we'd like until midnight December 31st. http://shortyawards.com/

[2] The haiku is derived from a previous form of poetry in vogue in Japan between the ninth and twelfth centuries, known as tanka. It had 3 lines, 5-7-5 syllables, treated religious themes or attached to the court. In the nineteenth century, the master Masaoka Shiki gave it a name: haiku (by the addition of words haiku and hokku). Matsuo Basho (1644-1694), was considered the first major Japanese haiku poet, was born samurai and adopted simplicity both in life and in poetic creation.

[3] http://www.c4lpt.co.uk/recommended/

educational purposes. One is that these tools are available to any user, teacher or student, who can use them without having to depend on business or education tools systems, which are often unwieldy, uncomfortable and unfriendly.

Based on the data available from the Centre for Learning & Performance Technologies that has published its list "Top 100 Tools for Learning", for the third consecutive year, it appears that Web 2.0 tools and open source applications are the trainers' favourites, and that nine out of ten tools used in teaching are open source (Castaño, 2009). The term cloud computing continues to gain the users' enthusiasm.

A report published by Morgan Stanley[4] on media consumption by adolescents, states: "teenagers do not use Twitter." A survey recently released by Nielsen Wire[5] data, from a representative sample of U.S. users of Twitter in different age groups, reinforces the idea that young people do not use this tool. The information available shows that only 16% of young people under 24 were using Twitter, it is the age group 24-54 who most use the service (64%), and those over 55 years old count for 20%. The data presented in this study_shows the popularity of the tool between users over 25. One explanation may be that youths use other social networks like Facebook or Hi5 which have an enormous capacity for sharing all kind of multimedia content, Twitter appears to them to have great limitations. It seems that young people, familiar with communicating quickly via SMS for free through their mobile phones, do not see Twitter as a useful service. When we look at the Nielsen Wire data, and at different generations, it seems that enthusiasm for Twitter is more for the Generation X (30 years old) and even the Baby Boomers (over 40 years old) than for the Net Generation. We think that teachers will have to show them its potential through curricular activities planned inside or outside the classroom, like our experiment.

Despite all the euphoria surrounding Twitter, Hart (2009), according to Gartner's Hype Cycle[6], believes that Twitter is entering the "Trough of disillusionment", however, many educators still seek to understand Twitter and use it in the classroom. Allowing each user to write what they are doing at the moment, means that not all messages posted on Twitter are relevant. A study by Pear Analytics[7] revealed a prevalence of tweets[8] or messages without relevant content.

For some authors, Twitter is an application for "narcissistis and egocentrics" (Serrano, 2009), for others it is an educational tool with potential (Cole, 2009; McNeil, 2009; Walker, 2009; Barrett, 2008). As a tool for teaching and learning (Barrett, 2008), teachers can use it in the classroom, to plan activities, conduct tutorials, create discussion forums, share addresses and launch questions. In turn, students can use Twitter to interact with each other anywhere in the world, learn about academic subjects, learn languages, and work in groups on different subjects. Wheeler (2009) presents the top ten uses of Twitter in education and encourages its use as a learning tool, both with computers or via mobile technologies. Among the listed uses we highlight the following:

[4] http://www.guardian.co.uk/business/morganstanley

[5] http://blog.nielsen.com/nielsenwire/online_mobile/teens-dont-tweet-twitters-growth-not-fueled-by-youth/

[6] http://www.gartner.com/DisplayDocument?ref=g_search&id=1108412&subref=simplesearch.

[7] http://www.pearanalytics.com/wp-content/uploads/2009/08/Twitter-Study-August-2009.pdf.

[8] A tweet is a post or status update on Twitter. Twitter only allows messages of 140 characters or less, "tweet" is as much a play on the size of the message as it is on the audible similarity to Twitter (Nations, 2009).

- inform students about changes of different kinds
- ask students to read an article or text or present a brief summary or keywords
- sharing some addresses targeted to pupils' tasks
- follow a famous person and document their work.

Wheeler also suggests the use of Twitter to make a brief summary of a document or achapter in just 140 characters, or to write micro stories, with only 20 words or to abbreviate an interesting story. This forces the user to emphasise the essentials, to be brief, and highlight in a schematic way the most noteworthly information. From the educacional point of view, Twitter can help develop important skills by encouraging people to write and develop their ability to summarise.

These are just some examples of how to use a small and simple tool the use of which has been growing exponentially since 2007. For some it has been the "buzz-word of microblogging". However, the importance of Twitter is not centred only on the application itself but also on the thousands of add-ons[9] that have been developed by programmers to give it more functionality.

Currently there is relatively little academic research on the use and impact of Twitter and only a few texts or communications have been published in conferences (McNeil, 2009). According to McNeil much of the work produced focuses on five categories:

- definitions of Twitter
- data on the growth and geographical distribution of Twitter users
- users' behaviour categorisation
- reflections on Twitter's use for education
- analysis of how Twitter works as back-channel.

Recently has been some discussion in the blogosphere about the use of Twitter.

2.1. Twitter: A Social Network

There are many social networks, some with more or fewer services, but among the most currently in use are Twitter and Facebook, with more than 65 million users (McGiboney, 2009). Some purists could say that Twitter is not really a social network, since it is a simple microblogging system not a multimedia system (Parsel, 2008). Facebook and Myspace are spaces that users can customize with network contacts, photos, videos and applications of all kinds. However, in Twitter there are no videos and no photos, only a few text characters and

[9] These are some of the many services in addition to Twitter to users: sharing photos flickr: snaptweet, management of followers: Tweepler; search engine of microblogging: twingly.com / microblogsearch; the last followers: twitter100; content most commented: tweetmeme ; customers: twhirl; posty; Twittastic; TwitterFon (iphone); tweetdeck; program: Twuffer; visually explore: twittervision; create your own community: twingr; information of followers: FriendOrFollow; twitter-friends; graphic network of friends: tweetwheel; save links in del.ici.ous: twitchboard; keep track of conversations: Tweetree; extension firefox: Power Twitter, TwitterFox; TwitterBar; polls: twtpoll; geolocation: twtvite, BD Application: Twitdom; send long tweets: twtlong; travel organizer : twtTRIP, creator of e-books: tweetbook.

everything happens in real time. Twitter is what McNeil (2009) called backchannel[10], "the growing adoption of Twitter has led to Twitter-enabled backchannels – both "official ", or "quasi-official", and "unofficial"- becoming an increasingly common feature of many academic conferences all over the world". According to Parsel (2008) Twitter is a network taken to extremes and in a network the most interesting thing is the connection and relationship. For him Facebook is a destination, while Twitter is the pure connection, so it is necessarily a departure point that is. Twitter is where we start but not where we finish. Twitter is fast and immediate it's complementary to other media and by sitting between them fills spaces that so far unfilled.

3. Twitter in Educational Institutions

In a study named *Twitter in Higher Education: Trends and Usage Habits of Today's College Faculty,* conducted by Faculty Focus[11] on using Twitter in higher education about two thousand university graduates, professors, deans, department staff, administrative services and scientific body were interviewed. The conclusions were not encouraging, because over half of the respondents did not incorporate Twitter into their teaching practices and only 1/3 of respondents said they used it in some way in university practice. The reasons given by the majority of respondents for not using Twitter was that they perceived it to be of little educational relevance and they had some concern that it created impoverished writing skills by reducing writing to so few characters. Other reasons were lack of interest and time to use the application, lack of safety data and lack of motivation in using Twitter because they saw it as being another fad of the social network era.

For those who use Twitter, tweets published on the network are on current news and trends related to interests in the academic field. There is interaction between colleagues and experiments in its use in the classroom. In this study, teachers who use microblogging as a medium of instruction, showed enthusiasm for Twitter and indicated an improvement in personal involvement and students' participation in academic activities.

There are several educational institutions all over the world that have adopted Twitter as a communication tool within the educational community. The North American universities were the first to exploit its potential and for those who want to belong to Web 2.0, they need to explore the usefulness of Twitter. Each institution must now to maximise the use of Twitter to get closer to their educational communities.

The Open University of Catalonia periodically updates its Twitter account to report about seminars and conferences, participation in fairs and exhibitions and agreements with other institutions. The Universidad Nacional de Educación a Distancia (UNED) publishes news, awards, exhibitions and relevant information to the university community. The European University of Madrid diligently updates its Twitter channel by publishing information on training courses, competitions, blog listings, students' associations, and other information.

---

[10] "The backchannel is a term used to designate the digital communications space used to sustain primarily textual interactions alongside live spoken presentations delivered in a physical environment" McNeil (2009).

[11] http://www.facultyfocus.com/free-report/twitter-in-higher-education-usage-habits-and-trends-of-todays-college-faculty/.

In Portugal this social network has gained a lot of fans, including artists, journalists and politicians[12]. The University of Porto was one of the first Portuguese universities to enter Twitter in order to maintain contact with the educational community, receiving real-time news and information about institutional events. The University of Coimbra also has a Twitter account. The University of Minho opened the *RepositoriUM* account on Twitter and users can follow the release of new documents and can follow news about the Institution. There are other Portuguese higher education institutions in this social network.

There are teachers from other countries who use Twitter and find ways to use it with students according to Barrett (2008) "In my experience, and in the short time that I have used it, Twitter has quickly grown to play a major part in the way that I interact with fellow colleagues and professionals from around the world. In my classroom and with the children I teach it has been an exciting tool to utilise and support learning".

It is important that educational institutions at all levels understand the potential that social networks offer to the education system. Using Twitter is a way to reinvent school 2.0. Nevertheless, as stated by Walker (2009), the biggest challenge is to get Twitter unblocked in school networks. But with Twitter installed on any mobile device that is no longer a problem.

Cole (2009) presents twenty-five ways of how teachers can use Twitter, we present only some of them: to ask for help, to get recommendations for lesson plans, to find books and professional resources, to contact other professionals and even to host a reading club online.

Walker (2009) shows some reasons for teachers to use Twitter:

- Twitter allows easy access to a huge, ever-changing resource pool of addresses, ideas, opinions and resources selected by professionals around the world and it eventually helps to overcome professional isolation. It's like having a virtual staff room.
- Twitter makes it possible to find professionals around the world and to know what they are doing. From the educational point of view it is a true revelation.
- It's great for teachers to reflect on what they are doing in their schools and see what is going well and what needs to be improved. Teachers can share ideas, support, and challenge each other.
- It can become a workshop of ideas and a sounding board. Twitter is an excellent way to share ideas and get instant feedback. Its speed and immediacy means we can get a range of opinions and constructive criticism in a matter of minutes.
- Having to express ideas in so few characters requires a lot of discipline, and Twitter is a great tool for that.

The ideas of Cole and Walker can inspire teachers to use Twitter as another educational tool and students to build on their own knowledge.

[12] It is estimated that in Portugal are already over 10 thousand users, has sparked interest for persons of different areas. In TwitterPortugal account, created by the Portuguese journalist Paulo Querido, let users consult the list of Portuguese and Brazilian service users by activities' sector. See: http://twitterportugal.com/.

4. Study Description

This experiment was part of a larger study on the use of Web 2.0 tools in an educational context. We followed a methodology of qualitative research, with an option for a Case Study, in order to address issues arising from specific educational situations and in contexts (Yin, 1984).

The case study as a qualitative method aims to discover the “how” and “why” that characterises the object of study. Although it has limitations, because it can not be generalised or lack objectivity, the case study can help understand of the subject under study, when the phenomenon is not isolated from its context (Hatley, 1994, Yin, 1984) as our case is.

This study sought to analyse the integration of Twitter and mobile technologies (in this case, the mobile phone) into the process of teaching and learning Portuguese language and litterature and was developed during the 3rd term of the 2008/2009 academic year.

The capacity for analytical and summary skills is socially valued and school should help to develop it in the students. This made us look at Twitter as a tool that is capable of enhancing these skills by limiting writing to 140 characters. Moreover, with the students’ existing familiarity with sending and receiving SMS via mobile phone, it seemed appropriate to create learning activities that combined Twitter and mobile phones.

Figure A. Twitter MobileG Website

As noted by several authors (Cole, 2009; McNeil, 2009; Walker, 2009; Barrett, 2008), Twitter is a learning tool that can be used in different contexts. We can turn it into a tool for collaborative or individual learning. Under these assumptions we proposed several microblogging activities using Twitter inside and outside the classroom, to complement the Portuguese language classes that can be accessed on the class students' website [13]. We challenged the students to publish micro stories, haikus, poems, proverbs, micro biographies, anecdotes, thoughts, aphorisms, riddles, unusual news items, quizzes, stanzas and diaries. Each student chose one of these suggestions and over the months they published tweets on the class Twitter account (figure A) which had been created for this activity[14].

Throughout the week the students wrote and stored their tweets on their mobile phones and in the class they posted them on Twitter through the school's computers in the lab. The mobile phone became a complementary educational tool, because it had facilitated the writing of the tweets (like SMS) and allowed their storage. Thus, it permitted learning to take place beyond the context of the classroom, and it happened anywhere and anytime, even on the move.

Some students sent their tweets via SMS to the teachers' mobile phone for assessment and correction before publishing them on Twitter. The students who had an Internet connection at home, posted tweets directly from home. We observed a very positive involvement by the students in all activities, and some of them posted more entries than originally planned. The published tweets were discussed afterwards and were analysed in the classroom. This activity benefited students as reported in the data analysis presented below.

4.1. Sample

For convenience we deliberately selected one of our classes as the sample group. The benefit of this was that the group was always available when needed for the research and we had control over the deadlines. This study involved of 16 students aged between 16 and 20 (all females) from the 10th grade of a Vocational Education class, of the Escola Profissional de Braga.

4.2. Data Collection Instruments

To collect data for this study a questionnaire was developed entitled "The Use of Twitter as a Learning Tool". It was completed anonymously by students through the online tool "SurveyMonkey". According to Tuckman (2005), inquiry research is a specific type of investigation and is often used in the education field. For Hill and Hill (2005), it is relevant in empirical research, to better understand the subject under study, particularly when making analysis.

The questionnaire included questions in both open and closed formats. To indicate the level of agreement or disagreement of students' perceptions about Twitter and its educational value, we used a Likert scale with 5 options (which we converted into 3 options: disagreement, indecision and agreement). To identify the students' perception of the relevance of the educational work carried out on Twitter and Twitter's usefulness as a

[13] http://sites.google.com/site/aulasdeportuguesonline/modulo-4.
[14] http://twitter.com/mobileG

complement of the activities done in the classroom, we used a scale with 3 options: no, in part, yes. The remaining questions about easiness, challenge and usefulness were dichotomous type (negative / positive).

The questionnaire consisted of several topics with sub-categories. For this chapter we selected the items corresponding to the follow topics:

i) Students perceptions about Twitter
ii) The educational value of Twitter
iii) The educational relevance of tasks
iv) The usefulness of Twitter as a complement to classroom activities.

4.3. Data Analysis

4.3.1. Students Perceptions about Twitter

We questioned students about their perceptions about Twitter (Table I) and the majority of them (75%) agreed that the information on Twitter was presented clearly. Almost all students (94%) said they felt comfortable writing on Twitter and would like to continue to publish in this application. When asked if the information on Twitter is very limited because of the 140 characters restriction, the majority (69%) were indecisive. When asked if in the future they intended to create a Twitter account, 68% were undecided. Only one student, in the group, had a Twitter account, and used it before this experiment.

4.3.2. The educational value of Twitter

Regarding the educational value of Twitter (Table II), most students (94%) considered that Twitter is good for communication. When asked about whether Twitter can be used for homework, only 47% agreed and 40% were indecisive, perhaps because most of the students do not have Internet access at home and they were restricted to publishing tweets only at school. When asked about if writing on the Twitter helped them to consolidate knowledge, 56% agreed and 44% were undecided. These results showed that Twitter has a potential educational value that should be exploited.

4.3.3. Educational relevance of information published on Twitter

Although most of the general messages that appear on Twitter seem to be irrelevant, when we promote its use in an educational context we need to create tasks that are relevant from an educational point of view. Therefore, we asked the students about whether the information published on Twitter promoted their interest in writing, about the Portuguese language and litterature in school work and knowledge in general (Table III) and the majority (56%) of the students responded “yes”, while 50% indicated that the information published on Twitter was relevant to homework only partly. It is natural that students who chose only to publish thoughts of the day or proverbs on Twitter did not feel the need to analysee and summarise information in a few characters as much as those who opted for writing a diary, news, micro stories or biographies. We need to continue to use Twitter in an educational context with new and more challenging tasks to make it a cognitive tool (Jonassen, 2000).

Table I. Students' perception about Twitter (n = 16)

Itens	Disagreement		Indecision		Agreement	
	f	%	f	%	f	%
On Twitter information is presented clearly.	0	0	4	25	12	75
I felt comfortable writing on Twitter.	0	0	1	6	15	94
I like to post on Twitter.	0	0	1	6	15	94
On Twitter information is very limited.	1	6	11	69	4	25
In the future I intend to create an account on Twitter.	3	19	11	68	2	13

Table II. The educational value of Twitter (n = 16)

Itens	Disagreement		Indecision		Agreement	
	f	%	F	%	f	%
Twitter is good for communication.	1	6	0	0	15	94
Twitter can be used for homework.	2	13	6	40	8	47
Through writing on Twitter, I can consolidate knowledge.	0	0	7	44	9	56

Table III. The educational relevance of information published on Twitter (n = 16)

Itens	No		In part		Yes	
	f	%	f	%	f	%
Writing	1	6	6	38	9	56
Portuguese Subject	1	6	6	38	9	56
Homework	1	6	8	50	7	44
General knowledge	0	0	6	38	10	62

Table IV. The usefulness of Twitter as a complement of classroom activities (n = 16)

Itens	No		In part		Yes	
	f	%	f	%	F	%
To carefully analyse the searched information.	0	0	4	25	12	75
To learn and select the most relevant information.	0	0	1	6	15	94
To summarise the main text ideas.	0	0	2	13	14	87
To organize the ideas logically.	0	0	2	13	14	87
To summarise.	0	0	0	0	16	100

Most students (62%) have considered that the information they published increased their knowledge in general, because a great variety of themes were shared during this activity.

4.3.4. The usefulness of Twitter as a complement to classroom activities

When students were questioned about whether Twitter helped them to carefully analyse the information that they had researched, the majority (75%) answered positively and only

25% responded by partially. When we asked about whether Twitter helped them to learn how to select the most relevant information, the majority (94%) said it did. Regarding the fact that Twitter demands condensed text, the majority (87%) of the students agreed that they had to focus on the main ideas and logically organise them. We asked students if Twitter helped them summarise information and all of them said it did. These results are in accordance with the considerations written in the literature review and the educational potential that Twitter can have in the classroom.

After analysing the answers we can conclude that Twitter can become a great tool for learning and helping students to develop significant expertise in summarising their writing in research and in the production of information.

4.3.5. Students experience perceptions: Answers categorisation

We asked some open questions to better understand the students' perceptions about this experiment of using Twitter as a supplement to the curricular activities of Portuguese language and litterature. The data presented in Table V records the students' perceptions about the:

- i) Relevance of writing in 140 characters
- ii) The benefit to the writing process
- iii) Sharing and interaction
- iv) Twitter's educational potential.

As regarding the relevance of writing in 140 characters, six sub-categories were identified:

- *Condense*
- *Summarise*
- *Simplify*
- *Organise ideas*
- *Improve analysis*
- *Easier reading.*

Students mentioned that: "Twitter helps me to exercise the brain"; "I learn to summarise text because I never knew how to do that before"; "Twitter helps me to analyse the information better"; "Twitter requires us to write in few words"; "It is easier to reading Twitter"; "I can say all I want objectively without writing too much"; "I summarise and organise ideas much better", "Twitter helps to synthesise information and it is a fast way to look at the news", "with Twitter I learned how to organize my ideas clearly"; "with Twitter we have to summarise the most important information".

As regards to Twitter's benefit to the writing process, five sub-categories were identified:

- learn to summarise
- practice writing
- learn to select information
- enhancing the capacity for condense
- improve writing motivation

These are some of the students' perceptions: "I had to know how to reduce my writing"; "I practiced my writing"; "I had to select the most relevant information"; "I learn to choose the most important details"; "it helped me to increase my ability to summarise"; "I learned how to write with a very limited number of words and it helped me to summarise the content"; "I loved it and I was always writing on Twitter".

For the third category, sharing and interaction, we considered three sub-categories as follows:

- help to learn from others
- help to share information
- arouse curiosity.

Students' wrote: "I found it interesting to know my classmates' works"; "it stimulates me to learn and share more"; "it helped me to better understand contents"; "we read work from each other"; "I try to always be aware of all account updates".

Finally, three sub-categories were identified for Twitter's educational potential:

- it is a learning and communication tool
- develops the intellect
- it promotes discovery

Some of the students' responses may lead us to reflect about Twitter's educational potential: "It is a working tool for learning"; "I think it is an educational site that helps students in writing and brain development; "I think it is a good learning tool primarily to help to summarize and make very short sentences to write only the most important details"; "It is fantastic. I already used Twitter before being referred to it in the Portuguese classes and I love it. I can be updated with all the news at the exact time of their release; I can communicate with everybody, including the director of one of my favourite series. Twitter is undoubtedly one of the best tools of our time"; "I believe that using Twitter is a fun way to learn new things"; "I think it's very good for us to communicate or for our culture".

The students recognised that Twitter has educational potential. They felt that by using it they developed essential writing skills (analysis, condense, and summarising), improved cognitive skills, shared knowledge, interacted and communicated with classmates. The reduced number of words requires polishing the content and training linguistic skills in a way to improve form and content.

We asked students whether the activities mediated by Twitter had been easy, challenging and useful. Most students (75%) felt that they had been easy, 94% said they were challenging and 87% mentioned they were useful. We also placed two questions about the usefulness of the mobile phone in the experiment and in the classroom. Most students (87%) felt that the mobile was a useful tool and 94% mentioned that the mobile phone should be used in school as a learning tool.

Table V. Students perceptions about the experiment (n = 16)

Category	Sub-Categories
Relevance of writing in 140 characters	Condense Summarise Simplify Organise ideas Improve the analysis Easier reading
Benefit to the writing process	Learn to summarise Practice Writing Learn to select information Enhancing the capacity for condense Improve writing motivation
Interaction and sharing	Help to learn from others Help to share information Arouse curiosity
Twitter's educational potential	It is a learning and communication tool It develops the intellect It promotes discovery

5. CONCLUSION

This is our first experiment using Twitter in an educational context. The results were very encouraging to its use as a learning tool. The students recognised its educational value in different areas. Students' perceptions went in the same direction as the characteristics and potential of Twitter. We presented before on the theoretical considerations. From the point of view of the curriculum, this experiment enabled the development of significant proficiency in the Portuguese subject: writing, condense, summarising, organisation of ideas, interaction and collaboration.

Although the development of writing skills, in particular the analysis of information and condense, is a school's teaching task, Twitter will allow the teacher to motivate students for these practices in a pleasant and participatory way, thereby ensuring that students have a more active role in building their own knowledge.

As a result of this experiment we believe that Twitter helps to develop important skills for the working world, for example: to be direct and objective, to use appropriate language, to provide useful information, to choose the right words; to hear and read the others, to collaborate and interact. Despite all opinions (good or bad) about Twitter, we believe that it can be very interesting from the educational point of view, if it is used wisely. Even with only a few characters available, this can be a benefit in helping to improve intellectual discipline and focus student thinking on the essential information. Mobile devices, such as mobile phones, can help to introduce a new educational concept – mobile learning, to educate the *net generation*, to improve learning skills and prepare students for the 21st century's challenges. According to Prensky (2001) "*our students have changed radically. Today's students are no longer the people our educational system was designed to teach*".

The research on the educational uses of Twitter and mobile technologies is still scarce so we must continue to invest in studies to assess its impact in education. The speed with which we can share links and external files can provide good experiences of learning, collaboration and cooperation. We have to accept our responsibility for change, only by doing this can we properly teach students in this digital era.

ACKNOWLEDGMENTS

This research is in part funded by CIEd, University of Minho.

REFERENCES

Barrett, T. (2008). *Twitter – A Teaching and Learning Tool*. http://tbarrett.edublogs.org/2008/03/29/twitter-a-teaching-and-learning-tool/.

Bell, J. (1989). *Doing your research project: a guide for the first-time researchers in education and social science*. Milton Keynes, England: Open University Press, 145.

Brondon, J. (2009). *Co-founder Biz Stone talks Twitter*. http://blogs.computerworld.com/biz_stone_interview/.

Castaño, C. (2009). 9 de cada 10 herramientas que utilizamos en el aprendizaje son libres. http://www.weblearner.info/2009/06/nueve-de-cada-diez-herraminetas-que.html.

Cole, S. (2009). 25 ways to teach with Twitter. In *Tech & Learning*. http://www.techlearning.com/article/20896/.

Esains, V. (2009). Herramientas 2.0 al servicio del aprendizaje. http://www.learningreview.com/tecnolog-para-e-learning-mainmenu-425/1733 herramientas-20-al-servicio-del-aprendizaje.

Frias, A. (2009). 'Twitter no es una red social, es para que todo el mundo sepa qué haces al momento'. In *elmundo.es*. http://www.elmundo.es/elmundo/2009/03/25/navegante/1237985543.html/.

Hart, J. (2009). *Twitter in the classroom: 10 useful resources*. http://janeknight.typepad.com/socialmedia/2009/08/twitter-in-the-classroom-10-useful-resources.html/.

Hartley, J. F. (1994). *Case studies in organizational research*. C., Cassell, & G. Symon, (sds.). Qualitative methods in organizational research: a practical guide. London: Sage, *253*, 208-229.

Hill, M. & Hill, A. (2005). *Investigação por questionário*. Lisboa: Edições Sílabo.

Iskold, A. (2007). Evolution of Communication: From Email to Twitter and Beyond. In Read Write Web. http://www.readwriteweb.com/archives/evolution_of_communication.php/.

Jansen, B. J., Zhang, M, Sobel, K. & Chowdury, A. (2009). Twitter Power: Tweets as Electronic Word of Mouth. *Journal of the American Society for Information Sciences and Technology*, *Vol. 9999*, 1-20.

Jonassen, D. H. (2000). *Computers as Mindtools for Schools: engaging critical thinking*, (2nd Ed.). Upper Saddle River: Prentice Hall.

Lemos, A. (2004). Cibercultura e Mobilidade: a Era da Conexão. Razan y palavra, nº 41, http://www.cem.itesm.mx/dacs/publicaciones/logos/anteriores/n41/alemos.html/.

Lévy, P. (2002). *L'intelligence collective et ses objets*. http://biblioweb. samizdat.net/article42.html/.

Martin, D. (2009). Twitter Quitters Post Roadblock to Long-Term Growth. In Nielsen Wire. http://blog.nielsen.com/nielsenwire/online_mobile/twitter-quitters-post-roadblock-to-long-term-growth/.

McGiboney, M. (2009). *Twitter's Tweet Smell Of Success*. Nielsen Wire. http://blog.nielsen.com/nielsenwire/online_mobile/twitters-tweet-smell-of-success/.

McNeil, T. (2009). *More than just passing notes in class? The Twitter-enabled backchannel.* http://www.scribd.com/doc/16287533/More-than-just-passing-notes-in-class-The-Twitterenabled-backchannel?autodown=pdf/ .

Nations, D. (2009). What is a Tweet? In *About.com: Web Trends*. http://webtrends.about.com/od/glossary/g/what-is-a-tweet.htm/ .

Oksman, V. & Rautiainen, P. (2003). Perhaps It Is a Body Part. How the Mobile Phone Became an Organic Part of the Everyday Lives of Children and Adolescents. J. E. Katz (Eds.), Machines that Become Us: The Social Context of Personal Communication Technology. New Brunswick: Transaction Publishers, 293-310.

Parselis, M. (2008). Se trata de la red al extremo. http://www.blog2.com.ar/index.php/2008/05/15/se-trata-de-la-red-al-extremo/.

Prensky, M. (2001). Digital Natives, Digital Immigrants. http://www.marcprensky.com/writing/Prensky%20-%20Digital%20Natives,%20Digital%20Immigrants%20-%20Part1.pdf.

Querido, P. (2009). Twitter: prós e contras em 140 caracteres (ou menos). In Cibercidadania – Expresso.http://aeiou.expresso.pt/gen.pl?sid=ex.sections/23475&num=10&page=&npages=50/.

Santaella, L. (2008). *A ecologia pluralista das midias locativas*. http://www.thefreelibrary.com/A+ecologia+pluralista+das+midias+locativas.-a0197040696/.

Serrano, R. (2009). Twitter una herramienta más en el aula. http://formacion-espana.universiablogs.net/2009/04/14/twitter-una-herramienta-mas-en-el-aula/.

Souza e Silva, A. (2006). Do ciber ao híbrido: Tecnologias móveis como interfaces de espaços híbridos. In: D. C. Araújo (Eds.), Imagem (Ir) realidade. Comunicação e cibermídia, Porto Alegre: Sulinas, 21-51.

Tuckman, B. (2005). *Manual de Investigação em educação*. Lisboa: Fundação Calouste Gulbenkian.

Walker, l. (2009). Nine great reasons why teachers should use Twitter. http://mrslwalker.com/index.php/2009/03/29/nine-great-reasons-why-teachers-should-use-twitter/.

Wheeler, S. (2009). *Teaching with Twitter*. http://steve-wheeler.blogspot.com/2009/01/teaching-with-twitter.html/.

Yin, R. K. (1984). Case study research: Design and methods. Newbury Park, CA: Sage.

York, D. (2007). The 10 ways I learned to use Twitter in 2007... (aka Why and How I use Twitter). http://www.disruptiveconversations.com/2007/12/the-10-ways-i-l.html.

Zago, G. S. (2008). *O Twitter como suporte para produção e difusão de conteúdos jornalísticos*. http://www.scribd.com/doc/5887184/O-Twitter-como-suporte-para producao-e-difusao-de-conteudos-jornalisticos.

In: Internet Issues: Blogging, the Digital Divide and Digital… ISBN: 978-1-60741-824-5
Editor: Cassie M. Evans, pp. 215-220

Expert Commentary

DISABILITY AND THE DIGITAL DIVIDE: GAPS AND FUTURE DIRECTIONS

Sally Lindsay*
Bloorview Kids Rehab and The University of Toronto, Toronto, Ontario, Canada

Access to information technology is no longer considered a luxury but rather, a valuable resource and an important survival tool to fully participating in society (Goggin and Newell 2007; Lindsay et al. 2007; 2008a; 2009b; Mirza et al. 2006). Although access to information technology (IT) and computer ownership has quickly increased, many people still lack access to these resources (Mirza et al. 2006). Such a digital divide between the 'haves' and the 'have-nots' highlights how marginalized groups have less access to information technology. Indeed, disability is often an overlooked aspect of the digital divide and should be taken into consideration when exploring digital inequality (Adam and Kreps 2006; Arrigo 2005; Dobransky and Hargittai 2006; Mirza et al. 2006). It is vital that information technology become more accessible for people with disabilities because it can play an important role in providing information, rehabilitation and social inclusion (Arrigo 2005).

Many studies have shown the benefits of using the Internet, especially for those with health problems or disabilities including: increased information and social support (Anderson et al. 1995; Dobransky and Hargittai 2006; Finn 1999; Guo et al. 2005; Ikemba et al 2002; Lindsay et al. 2008; 2009; Norris 2001; Swartz 1996; Zaidman-Zait and Jamieson 2007), improved frequency and quality of social interaction, improved independence and self-determination (Cook et al. 2005; Mirza 2006), improved health related quality of life, decreased isolation, reduced health care utilization (Drainoni et al. 2004; Lorig et al. 2006; Magnusson et al. 2004;), and reduced social barriers in the physical and social environment (Bradley and Poppen 2003; Guo et al. 2005).

Evidence also suggests that information technology can provide people with disabilities with a new set of tools for social participation and personal development enabling them to

* Sally Lindsay, PhD, Scientist, Bloorview Research Institute, Bloorview Kids Rehab, Assistant Professor, Dalla Lana School of Public Health, University of Toronto, 150 Kilgour Road, Toronto, Ontario Canada M4G 1R8, Email: slindsay@bloorview.ca

find new contacts with others in a similar situation (Anderberg 2007). Online communication can allow people with disabilities to interact with others to an extent that may not be possible offline such as removing one's disability status from the forefront of interaction (Bowker and Tuffin 2002; Davidson 2008; Dobransky and Hargittai 2006; Guo et al. 2005). There is little doubt that there are many benefits of using information technology, particularly for people with disabilities; however, technology needs to be accessible in order to tap into such benefits.

Many barriers remain in accessing information technology for people with disabilities despite government policies and standards concerning the usability, accessibility and universal design of the Internet (Arrigo 2005). As such, people with disabilities are less likely than the population as a whole to use computers or the Internet (Mirza et al. 2006). Some estimates suggest that approximately 25% of people with disabilities use the Internet compared to 56% of people without disabilities (Kaye 2003).

Cost is often a major barrier for people with disabilities accessing the Internet because they often need assistive devices and adapted technology, which can be expensive (Hollier and Murray 2004; Mann et al. 2005). This is further complicated by the fact that people with disabilities often have lower incomes than people without disabilities and thus, may not be able to afford an adapted computer (D'Aubin 2007; Dobransky et al. 2006).

Furthermore, inappropriately designed technology is another key barrier to accessing the Internet for people with disabilities. Hardware or software may not be configured to allow people with disabilities to use the Internet effectively (Dobransky et al. 2006). Despite international guidelines for designing accessible websites, most sites are not compatible with the assistive technology that visually impaired computer users typically need to convert text to speech output (Disability Rights Commission 2004; D'Aubin 2007). For instance, people who are deaf or who have hearing impairments often cannot access video clips on the Internet because they rarely provide closed captioning (D'Aubin 2007).

A main hurdle for people with disabilities is that assistive technology is often reactive in design; thus, by the time accommodations are made technology has often changed (Dobransky et al. 2006). Some researchers argue that as multimedia content becomes more widespread people with disabilities will encounter even more barriers (Bruyere et al. 2006; Dobransky et al. 2006). For example, one study examining the accessibility of websites for people with disabilities found that less than 20% of websites met even the most basic accessibility standards across a spectrum of disabilities including hearing, motor and sight impairment (Adam and Kreps 2006; Disability Rights Commission 2004). This is particularly concerning given the increasing use of the Internet for information, social, cultural and political participation. Lacking access to information technology can have serious implications for social inclusion—possibly even reinforcing the marginalized status of people with disabilities (Adam and Kreps 2006; Cuneo 2002; Drainoni et al. 2004; Gogin and Newell 2003; Kaye 2003).

Many technologies are intimately associated with our bodies and have a complex relationship to impairment and disability (Goggin and Newell 2007). Thus, the introduction of new technologies can create new forms of exclusion for people with disabilities. There is little doubt that people with disabilities have the knowledge, skills, and willingness to use information technology (Egan et al. 2005), however, they often encounter barriers in accessing it (Hollier and Murray 2006).

In summary, the digital divide literature has tended to largely ignore disability (Adam and Kreps 2006). For example, several major works in the field offer little or no discussion of disability (Loader 1998; Marshall et al. 2003; Norris 2001; Servon 2002). More work is needed, therefore, to shed light on the barriers and inequalities that persist in accessing information technology among people with disabilities. First, a more proactive approach is needed where technology is created to be accessible in the first place (Stephandis and Emiliani 1999). Second, more research should focus on the use of information technology among people with disabilities. Few studies have examined IT usage patterns among people with disabilities and the particular barriers that they may encounter in doing so (Drainoni et al. 2004). For instance, what are the physical, social and environmental barriers that people with disabilities encounter while attempting to use the Internet? Furthermore, most of the focus has been on adults while less attention has been paid to children's perspectives and the barriers that they encounter when going online.

Third, many research studies focus on IT use amongst people with disabilities who already own computers and have access to the Internet, but what about people who do not own computers? What are the characteristics of people with disabilities who would like to use information technology but have difficulties accessing it? Further, how do people with disabilities access the Internet in public spaces; and do they encounter barriers in doing so?

Fourth, future studies should focus on what is happening at the juncture of disability, accessibility and technology (Goggin and Newell 2007), particularly how the social shaping of the Internet influences how disability and accessibility are constructed within these spaces. Finally, there is a greater need for policy to focus on accessibility and affordability of information technologies to help close this digital divide and to increase social inclusion for people with disabilities (Dobransky et al. 2006; Kaye 2003; Smith 2009; Smith et al. 2009).

Finally, relatively little is known about IT use by type of disability. People with disabilities should not be viewed as a homogenous group because there are many differences between them (e.g., physical versus intellectual disabilities), which are often not often considered. Thus more research is needed to explore the accessibility to technology for people with disabilities and especially how this varies by the type of condition, age, socio-economic status, gender, ethnocultural status, geographical location and so forth (Annable et al. 2007). In sum, many people with disabilities are still unable to effectively use information technology—excluding them from full participation in society (D'Aubin 2007). More effort is needed to increase the accesssiblity of information technology for people with disabilities.

REFERENCES

Adam, A., & Kreps, D. (2006). Web Accessibility: A digital divide for disabled people? In *Social inclusion: Societal and organizational implications for information systems.* International Federation for Information Processing. Boston: Springer, pp. 217-228.

Anderberg, P. (2007). Peer assistance for personal assistance: analysis of online discussions about personal assistance from a Swedish web forum for disabled people. *Disability & Society,* 22(3), 251-265.

Anderson, R., Bikson, T., Law, S., & Mitchell, B. (1995). *Universal access to e-mail: feasibility and societal implications.* Rand, Santa Monica, CA.

Annable, G., Goggin, G., Stienstra, D. (2007). Accessibility, disability and Inclusion in Information Technologies: an introduction. *The Information Society,* 23, 145-147.

Arrigo, M. (2005). Online distance education and visual disability. *Methods and Technologies for Learning*, 577-582.

Bowker, N., & Tuffin, K. (2002). Disability discourses for online identities. *Disability and Society*, 17, 327-344.

Bradley, N. & Poppen, W. (2003). Assistive technology, computers and the internet may decrease sense of isolation for homebound elderly and disabled persons. *Technology and Disability*, 15, 19-25.

Bruyere, S., Erickson, W., & VanLooy, S. (2006). Information technology accessibility: implications for employment of people with disabilities. *Work,* 27, 397-405.

Cook, J., Fitzgibbon, G., Batteiger, D., Grey, D., Caras, S., Dansky, H., & Priester, F. (2005). Information technology attitudes and behaviors among individuals with psychiatric disabilities who use the Internet: results of a web-based survey. *Disability Studies Quarterly*, 25(2).

Cuneo, C. (2002). Globalized and localized digital divides along the information highway: A fragile synthesis across bridges, ramps, cloverleaves and ladders. *33rd Annual Sorokin Lecture,* January 31.

D'Aubin, A. (2007). Working for barrier removal in the ICT area: creating a more accessible and inclusive Canada. *The Information Society,* 23, 193-201.

Davidson, J. (2008). Autistic culture online: virtual communication and cultural expression on the spectrum. *Social & Cultural Geography,* 9(7), 791-806.

Disability Rights Commission (2004). *The web: Access and inclusion for disabled people.* London: The Stationary Office.

Disability Rights Commission. (2004). *The web: Access and inclusion for disabled people.* Disability Rights Commission, London.
http://www.drc-gb.org/library/webaccessibility.asp

Dobransky, K., & Hargittai, E. (2006). The disability divide in internet access and use. *Information, Communication & Society,* 9(3), 313-334.

Drainoni, M., Houlihan, B., Williams, S., Verandi, M. et al. (2004). Patterns of Internet use by persons with spinal cord injuries and relationship to health-related quality of life. *Archives of Physical and Medical Rehabilitation*, 85, 1872-1879.

Egan, J., Worrall, L., & Oxenham, D. (2005). An Internet training intervention for people with traumatic brain injury: barriers and outcomes. *Brain Injury*, 19(8), 555-568.

Finn, J. (1999). An exploration of helping processes in an online self-help group focusing on issues of disability. *Health and Social Work,* 24, 220-231.

Goggin, G. & Newell, C. (2003). *Digital disability: the social construction of disability in new media.* Rowman and Littlefield, Lanham, MD.

Goggin, G. & Newell, C. (2007). The business of digital disability. *The Information Society,* 23, 159-168.

Guo, B., Bricout, J., & Huang, J. (2005). A common open space or a digital divide? A social model perspective on the online disability community in China. *Disability & Society,* 20(1), 49-66.

Hollier, S. & Murray, I. (2004). The evolution of e-inclusion—Technology in education for the vision impaired. *History of Computing in Education,* 145, 123-131.

Ikemba, C., Kozinetz, C., Feltes, T., & Fraser, C., et al. (2002). Internet use in families with children requiring cardiac surgery for congenital heart disease. *Pediatrics,* 109, 419-422.

Kaye, S. (2003). Disability and the digital divide: accounting for lower levels of computer and internet usage among people with disabilities. *131st Annual Meeting of APHA*, November 15-19.

Lindsay, S., Smith, S., Bell F. & Bellaby, P. (2007) Tackling the digital divide: exploring the impact of ICT on managing heart conditions. *Journal of Information, Communication & Society,* 10(1), 95-114.

Lindsay, S. & Smith, S. (2008). The role of social capital in research participation. *Reinventing Trust: Discourses on Social Capital.* (Chapter 13) In K. Gupta, G. Svendsen, P. Maiti (Editors) New Delhi: Atlantic Books.

Lindsay S., Bellaby, P., Smith, S. & Baker, R. (2008a). Enabling healthy choices: Is information and communication technology the highway to health improvement? *Health: An Interdisciplinary Journal in the Study of Health and Illness,* 12(3), 313-331.

Lindsay, S., Smith, S. and Bellaby, P. (2008b). Can informal e-learning bridge the digital divide? *Social Policy and Society,* 7(3), 1-12.

Lindsay, S., Smith, S., Baker, R. & Bellaby, P. (2009) The health impact of an online heart disease support group: A comparison of moderated vs. unmoderated support. *Health Education Research*, 24(4): 646-654.

Loader, B. (1998). *Cyberspace Divide: Equality, agency and policy in the information society*. London: Routledge.

Lorig, K., Ritter, P., Laurent, D. & Plant, K. (2006). Internet-based chronic disease self-management: a randomized trial. *Medical Care,* 44(11), 964-971.

Magnusson, L., Hanson, E., & Borg, M. (2004). A literature review study of information and communication technology as a support for frail older people living at home and their family carers. *Technology and Disability,* 17, 1-9.

Mann, W., Belchoir, P., Tomita, M., & Kemp, B. (2005). Computer use by middle-aged and older adults with disabilities. *Technology and Disability*, 17, 1-9.

Marshall, S., Taylor, & You, X. (2003). *Closing the digital divide: transforming regional economies and communities with information technology*. Westport, CT: Praeger.

Mirza, M., Anadan, N., Madnick, F., & Hammel, J. (2006). A participatory program evaluation of a systems change program to improve access to information technology by people with disabilities. *Disability and Rehabilitation*, 28(19), 1185-1199.

Norris, P. (2001). *Digital divide: Civic engagement, information poverty, and the internet worldwide.* Cambridge, UK: Cambridge University Press.

Servon, L. (2002). *Bridging the digital divide: technology, community and public policy.* Malden, MA. Blackwell.

Smith, S. (2009). From inclusive spaces to inclusionary texts: how eParticipation can help overcome social exclusion. In E. Ferro, Y. Dwivedi, J. Gil-Garcia and M. Williams (eds). *Handbook of research on overcoming digital divides,* Chapter 29. IGI Publishers.

Smith, S., Lindsay, S., Bellaby. (2010) Social inclusion at different scales in the urban environment: locating the community to empower. *Urban Studies,* 47(7): 1439-1457.

Stephandis, C. & Emiliani, P. (1999). Connecting to the information society: a European perspective. *Technology and Disability,* 10, 21-44.

Swartz, E. (1996). *Net activism: how citizens use the internet.* Songline studios, Sebastopol, CA.

Zaidman-Zait, & Jamieson, J. (2007). Providing web-based support for families of infants and young children with established disabilities. *Infants and Young Children,* 20(1), 11-25.

INDEX

A

academics, 54, 57, 59, 69, 79, 184
accessibility, 179, 184, 216, 217, 218
accountability, 6, 8
activism, 28, 50, 57, 66, 220
adiposity, 144
administrators, 77, 79, 80
adolescents, 164, 178, 184, 185, 202
advertising, 62, 77
aesthetics, 5
aetiology, 66, 70
age, x, 61, 62, 63, 73, 76, 118, 123, 131, 134, 135, 138, 139, 140, 141, 149, 152, 153, 154, 156, 158, 178, 179, 184, 202, 217
aggregation, 101, 108
algorithm, 107, 113, 135, 192, 193
alternatives, 134, 135
analytical framework, 54, 69
animations, xii, 189, 190, 193, 194
anxiety, 163, 166, 167
applied research, 187
artificial intelligence, 65
assessment, 63, 207
assistive technology, 216
assumptions, xii, 25, 101, 199, 207
athletes, 77, 80
attacks, 162, 164
attitudes, 12, 16, 62, 63, 64, 67, 85, 120, 148, 152, 159, 161, 178, 218
attractiveness, 31, 182
authenticity, 78
authority, 28, 36, 41, 51, 54, 55, 87
autonomy, 173
awareness, 8, 11, 18, 21, 31

B

background, 8, 18, 107, 120, 149
background noise, 8
barriers, 97, 119, 169, 172, 184, 186, 215, 216, 217, 218
behavior, xi, 80, 106, 110, 114, 115, 145, 151, 177, 185, 186, 187
beliefs, 24, 61, 62, 63, 64, 152, 178
bioethics, 59, 167, 175
black market, 30
blocks, 156
blogger, viii, 5, 17, 21, 29, 32, 33, 34, 53, 55, 56, 57, 59, 64, 65, 66, 67, 84, 151
bonding, 127, 132, 133
brain, 210, 211
broadband, 122, 124, 132, 138
browser, 72, 73, 100, 196
browsing, 86, 88, 89, 103, 200
buffer, 60
burning, 167

C

campaigns, 32, 77, 154
candidates, 32, 154
cardiac surgery, 219
case study, viii, 27, 48, 116, 206
cast, 149, 182, 185
catalyst, 120
catastrophes, 168
categorization, 84, 88, 89, 185
category d, 89
cell, 35, 36, 40, 45, 73, 79, 178, 185
cell phones, 79, 178, 185
Census, 103

channels, 30, 35, 61
chaos, 78
Chief Justice, 155
child development, 25
children, x, xi, 4, 69, 118, 122, 126, 127, 134, 177, 178, 184, 185, 186, 187, 205, 217, 219, 220
chimera, 48
citizenship, 162, 163, 164
civil liberties, 149, 151, 160, 163, 164
civil rights, 149, 153, 155
clarity, 7, 8
classes, 75, 77, 80, 121, 207, 211
classification, viii, 53, 54, 60, 67, 69, 84, 85, 93
classroom, vii, ix, xii, 1, 2, 7, 10, 12, 15, 17, 18, 23, 71, 74, 115, 199, 201, 202, 204, 205, 207, 208, 209, 210, 211, 213
classrooms, vii, 1, 10
cleaning, 106, 107
closure, 164, 166
cluster analysis, 28, 110, 116
cluster model, 110
clustering, ix, 95, 108, 109, 110, 112, 113, 115
clusters, viii, 27, 36, 37, 40, 42, 43, 44, 45, 46, 47, 109, 110, 111, 113
CMC, viii, 53, 54, 60
CNN, 31
coaches, 78, 80
codes, 100
coding, 29
cognitive ability, 160
cognitive function, 141
cognitive process, 9
cognitive tool, 200, 208
coherence, 45, 46, 47
collaboration, 14, 56, 59, 69, 72, 73, 74, 76, 77, 85, 212
college students, 72
colleges, 74, 76, 77, 78, 80
commodity, 195
communication strategies, 144
communication technologies, 129, 199
communism, 153
Communist Party, 36, 43
community, viii, 2, 5, 6, 9, 13, 16, 19, 24, 29, 30, 34, 38, 41, 49, 51, 53, 55, 56, 57, 58, 59, 60, 62, 65, 67, 73, 74, 76, 78, 80, 84, 85, 87, 90, 91, 92, 112, 118, 120, 122, 151, 167, 170, 173, 191, 192, 196, 201, 203, 204, 205, 218, 219
competence, 65, 201
complement, xi, xii, 178, 199, 201, 207, 208, 209
complexity, viii, 7, 53, 54, 60, 67, 80, 108, 194
components, vii, 1, 39, 160, 181, 192
composition, 67, 190
comprehension, 141
computer skills, 79, 97
computer technology, 170, 187
computer use, 129, 216
computing, 103, 115, 196, 202
concentrates, 86, 89
configuration, 35
conflict, 125, 161
conformity, 164
connectivity, viii, 27, 28, 36, 42, 45, 46, 47
connectivity patterns, viii, 27, 36, 42
consensus, 61, 80, 148, 149
consent, 114, 138
construction, xii, 2, 15, 29, 69, 108, 189, 196
consumers, 75, 113, 120
consumption, 185, 202
content analysis, 86, 87, 88, 89, 90, 92
contradiction, vii, 27, 33
control, x, 12, 13, 29, 79, 117, 122, 124, 142, 151, 173, 207
control group, 122
convergence, 162, 200
coronary heart disease, 121, 128
correlation, x, 43, 100, 131, 138
correlation analysis, x, 131
correlations, 135
creationism, 66
creativity, xii, 74, 199, 201
credibility, 48, 50, 79, 162
credit, 57, 74
critical thinking, vii, 1, 2, 213
criticism, 17, 205
cues, 180, 187
cultural practices, 11
cultural values, 6
culture, 22, 55, 64, 92, 187, 211, 218
curiosity, 12, 211, 212
currency, 29
current limit, ix, 95
curricula, 81
curriculum, 13, 94, 98, 212
customers, 30, 203

D

data analysis, vii, 1, 19, 207
data collection, 179, 185
data mining, ix, 95, 96, 106, 107, 109, 113, 114, 115
data set, x, 19, 20, 40, 131
database, 81, 82, 90, 98, 100, 106, 107, 194
deaths, 166, 174, 184
decision making, 142
decisions, 79, 107, 109, 123, 147, 167, 169, 172

definition, 8, 33, 84
delivery, 61, 166, 169, 174
democracy, x, 44, 147, 148, 149, 151, 152, 153, 160, 163, 164, 171
demographic characteristics, x, 131, 138, 142
demographic factors, x, 147, 148
demographics, 74, 77, 82, 110, 154, 156, 180, 190
density, viii, 28, 37, 38
dependent variable, 135, 137, 140, 156
depression, 65
deprivation, 121
designers, 96, 113
detection, 193
developed nations, 167
deviation, 37, 38
diabetes, 185
diet, 172, 184
dietary habits, 185
diffusion, x, 117
digital communication, 204
digital divide, iv, vii, x, xi, xii, 117, 118, 120, 121, 128, 131, 132, 134, 143, 165, 170, 171, 175, 215, 217, 218, 219
digital technologies, 3, 120
directives, 170, 172
disability, xii, 118, 119, 120, 171, 173, 215, 216, 217, 218
discipline, viii, 35, 53, 54, 59, 60, 63, 64, 65, 66, 67, 205, 212
disclosure, 57, 63, 64, 79
discourse, viii, 31, 48, 53, 54, 56, 60, 61, 67, 69, 163, 169
discrimination, 118
displacement, 133, 185
disposition, 121
disseminate, 48, 56, 91, 172
distance education, 218
distribution, 28, 31, 34, 37, 38, 41, 42, 59, 101, 102, 112, 203
divergence, 45
diversity, 185
division, xi, 165
dominance, 141

E

e-commerce, 129
economic crisis, 61
economic downturn, 132
economic status, x, 131, 136, 145, 217
education, 1, 2, 3, 22, 23, 24, 25, 27, 48, 50, 51, 52, 56, 69, 71, 76, 81, 82, 86, 88, 89, 115, 127, 128, 134, 136, 137, 140, 143, 145, 152, 157, 161, 162, 172, 176, 195, 199, 204, 207, 218, 219
educational attainment, x, 131, 134, 135, 138, 140
educational process, 79
educational qualifications, 118
educational research, 113
educational software, 122
educational system, 106, 212
egalitarianism, 143
ego, 84
e-Government, 128
elderly, 118, 218
e-learning, 91, 128, 213, 219
election, 31, 32, 91, 160, 161
elementary teachers, 25
elucidation, 7
email, 3, 4, 72, 77, 79, 82, 119, 124, 126, 200
emotions, 22, 85
employees, 30, 80
employment, 126, 218
empowerment, 128
encouragement, 3, 127
endorsements, 154
energy, 86
engagement, vii, 1, 2, 4, 13, 14, 15, 18, 20, 77, 81, 82, 129, 186, 219
enrollment, 78
enthusiasm, 202, 204
environment, 14, 19, 50, 59, 60, 74, 109, 114, 118, 122, 127, 133, 148, 151, 160, 161, 178, 219
epidemic, 167
epidemiology, 66
equity, xi, 165, 169
ethical issues, 166, 167, 169, 175
ethics, 122, 166, 174
ethnic groups, 74
ethnic minority, 179, 185, 186
ethnicity, 74, 180, 182
etiquette, 78
euphoria, 202
European Commission, 128, 145
evolution, 213, 218
examinations, 159
exclusion, 118, 121, 216
exercise, 63, 151, 159, 210
expertise, 55, 149, 152, 153, 160, 171, 210
exposure, x, 74, 147, 148, 149, 150, 151, 153, 155, 156, 157, 159, 160, 161, 162, 163, 164, 200
extraction, 106, 184, 190, 192

F

Facebook, ix, 71, 72, 73, 76, 77, 78, 79, 80, 81, 82, 169, 201, 202, 203
face-to-face interaction, 132, 150
family income, 103, 104, 105
family interactions, 120
family life, 127, 129
family members, x, 117, 120, 121, 122, 123, 124, 125, 126, 127
family relationships, 121
fear, 32, 79, 123, 149, 151, 166, 167
feedback, 21, 29, 30, 57, 58, 60, 75, 200, 205
feelings, 8, 11, 66, 84, 167
females, 135, 139, 140, 141, 182, 184, 207
films, 188
filters, ix, 29, 71, 72
floating, 168
focus groups, x, xi, 114, 117, 122, 177, 179, 180, 181, 182
focusing, ix, x, 83, 86, 87, 88, 90, 117, 118, 218
football, 80
framing, 121, 162
freedom, xi, 9, 21, 91, 147, 149, 160
fuel, 125
funding, 28, 32

G

gait, 185, 187
gay marriage, 161
gender, x, xi, 61, 62, 63, 67, 131, 133, 134, 135, 138, 139, 141, 153, 154, 156, 158, 160, 177, 178, 180, 182, 184, 186, 217
gender differences, xi, 133, 177, 179
gender effects, 160
general election, 32, 50
generalization, 108
generation, 22, 25, 75, 82, 121, 190, 199, 200, 212
genre, 49, 54, 56, 61, 69, 178, 182, 184, 185, 187
genres, viii, xi, 53, 54, 55, 177, 178, 179, 181, 184, 185
geolocation, 203
girls, 178, 182, 185
goals, 1, 19
google, 100, 115, 207
government, 36, 107, 119, 120, 129, 153, 169, 170, 196, 216
grades, 97, 181
graduate students, 55, 60
grains, 190, 191, 192, 193, 194, 195
grants, 36
graph, 35, 36, 39, 43, 47
gravity, 7, 9
group membership, 110
groups, viii, xii, 22, 28, 35, 36, 37, 40, 42, 47, 54, 58, 60, 61, 67, 76, 80, 91, 103, 109, 110, 113, 120, 122, 133, 141, 142, 148, 149, 150, 151, 152, 153, 159, 164, 170, 173, 174, 179, 180, 181, 182, 202, 215
growth, 22, 54, 82, 96, 101, 104, 105, 133, 147, 160, 178, 200, 202, 203, 214
growth rate, 104
guidelines, 216

H

hands, 4, 5, 98, 99
harm, 166, 167, 168, 171, 174, 175
health, x, xi, 59, 66, 117, 118, 119, 121, 122, 124, 128, 129, 131, 132, 133, 134, 135, 136, 137, 138, 141, 142, 143, 144, 145, 166, 169, 172, 173, 174, 177, 179, 185, 186, 215, 218, 219
health care, 129, 166, 174, 215
health education, 144
health information, 119, 132, 133, 141, 143, 144
health problems, 119, 215
health services, 133, 142
health status, 134
hearing impairment, 172, 216
heart disease, 121, 128
heterogeneity, 150, 161, 163
higher education, ix, x, 2, 3, 4, 52, 71, 76, 78, 80, 82, 131, 133, 139, 140, 141, 204, 205
Hispanics, 154
homework, 208, 209
homogeneity, 164
hopes, 123, 127
host, 72, 133, 205
household income, 29, 134, 154
households, 118, 126, 132, 138, 143, 170, 178
hub, 34, 58, 68, 91
human activity, 17
human development, 10
hybrid, viii, 23, 48, 53, 54, 55, 58, 68, 84, 92, 195
hypertext, 35, 86, 87
hypothesis, viii, 27, 37, 42, 47, 159

I

icon, 56
identification, viii, 27, 186, 193
identity, 21, 63, 64, 66, 79, 92
ideology, viii, 53, 148, 151, 156, 158, 160, 163

image, 15, 91, 99, 190, 191, 193, 194
imagery, 187
images, xii, 5, 7, 14, 75, 186, 189, 190, 194, 195
immersion, xi, 178, 185, 186
implementation, 114
implicit knowledge, 106
inclusion, xii, 66, 110, 128, 129, 196, 215, 216, 217, 218, 219
income, x, xi, 74, 103, 118, 125, 131, 134, 135, 138, 149, 152, 156, 158, 170, 172, 174, 177, 179, 184, 185
independent variable, 156
indexing, 194
indicators, 20, 90, 110, 111, 112, 113, 144
indirect effect, 158
individual action, 167
individual rights, 148, 166
industry, 30, 49, 50, 184, 186
inequality, xii, 172, 215
infancy, 80
infants, 220
infection, 168
infectious disease, 66, 166, 167, 174, 175
inferences, 114, 185
information exchange, 33, 61
information processing, 163
information sharing, 5, 201
information technology, vii, xii, 171, 215, 216, 217, 219
informed consent, 141
infrastructure, xi, 115, 165, 167, 168, 170, 173, 174, 175
inner world, 6, 7
innocence, 186
innovation, 24, 48
insight, vii, 120, 184
institutions, 54, 58, 59, 76, 78, 80, 102, 166, 204, 205
instruction, 25, 204
instructors, 75
instruments, 32, 48
integration, 106, 107, 115, 195, 206
intellect, 211, 212
intellectual disabilities, 217
intelligence, 85, 213
intentions, 114
interaction, x, 14, 15, 35, 53, 56, 61, 62, 64, 68, 75, 79, 115, 131, 135, 136, 137, 140, 142, 150, 160, 186, 200, 204, 210, 211, 212, 215, 216
interaction effect, 136
interactions, 5, 6, 11, 28, 35, 77, 93, 125, 132, 133, 134, 135, 204
interest groups, 72
interface, 85
intermediaries, 120
internalization, 149
internet, vii, ix, xi, 15, 21, 29, 33, 49, 50, 72, 83, 93, 128, 129, 134, 143, 165, 166, 168, 169, 170, 171, 172, 173, 174, 175, 218, 219, 220
interoperability, 99
interpersonal relationships, 80
intervention, x, 10, 127, 131, 132, 134, 139, 140, 142, 179, 185, 187, 218
interview, 129, 200, 213
intimacy, 133
intonation, 6
investment, 19, 119, 178
investors, 55
IP address, 102, 106
Iraq War, 162
isolation, 65, 166, 205, 215, 218

J

jobs, 76, 122
journalism, 30, 31, 48, 51, 52, 84, 92, 151
journalists, 30, 31, 55, 64, 205
judgment, 31
justice, xi, 165, 174
juvenile diabetes, 187

L

labeling, 62
labour, 121
land, 121
landscape, vii, 1
language, 2, 4, 5, 6, 7, 8, 9, 10, 11, 12, 13, 15, 17, 18, 20, 24, 61, 64, 85, 88, 89, 92, 143, 171, 172, 190, 206, 207, 208, 210, 212
language barrier, 171, 172
laptop, 3
learners, vii, 1, 3, 4, 8, 9, 10, 11, 13, 14, 15, 16, 17, 18, 19, 20, 21, 75, 82, 96, 120
learning, vii, ix, xii, 1, 2, 3, 4, 5, 8, 10, 11, 12, 13, 14, 15, 16, 17, 18, 19, 20, 21, 22, 23, 24, 25, 30, 50, 52, 75, 77, 92, 95, 96, 97, 98, 101, 109, 112, 113, 114, 115, 119, 120, 121, 125, 195, 196, 199, 200, 201, 202, 205, 206, 207, 210, 211, 212, 213
learning environment, 1, 10, 15, 19, 92, 113, 115
learning outcomes, vii, 1, 16, 17
learning skills, 212
legend, 193
leisure, 120, 128
lesson plan, 112, 205

Libertarian, 154
librarians, 48, 86, 87, 88, 89, 90, 91
life sciences, 59
lifestyle, 16, 121
limitation, 114, 136, 201
line, 9, 23, 29, 30, 40, 77, 114, 124, 133, 136, 196
links, viii, ix, 21, 28, 33, 34, 35, 36, 37, 38, 39, 40, 41, 42, 43, 44, 45, 46, 47, 48, 56, 59, 64, 71, 72, 77, 84, 86, 87, 97, 124, 141, 163, 203, 212
listening, 10, 14, 15, 31, 136, 181
literacy, 15, 68, 97, 100, 119, 129, 141, 144
local community, 2, 132
longitudinal study, 137
love, 64, 178, 186, 211

M

management, xii, 50, 56, 79, 85, 97, 98, 115, 121, 185, 195, 199, 203, 219
mapping, 35
market, 77
marketing, 31, 78, 81, 82, 84, 85, 105, 187, 201
marketing strategy, 105
marketplace, 152
markets, 91
mass media, 34, 163
material resources, 172
matrix, 35, 36, 38, 39, 40, 45, 111
meanings, 6, 10, 11, 19, 21
measures, 101, 109, 110, 114, 115, 152, 159, 160, 162, 166
media, ix, x, xi, xii, 11, 15, 28, 30, 31, 32, 33, 34, 48, 50, 55, 57, 63, 71, 72, 76, 78, 80, 81, 82, 84, 128, 129, 131, 134, 135, 136, 141, 149, 150, 151, 152, 153, 159, 160, 161, 162, 165, 167, 168, 169, 172, 175, 177, 185, 186, 188, 189, 190, 202, 204
median, 34, 103, 104, 105, 154
membership, 21, 59, 110
men, 121, 122, 123, 133, 139, 141, 153
mental imagery, 178
mental processes, 10
mentoring, 69
messages, xi, xii, 31, 61, 78, 84, 86, 141, 167, 169, 170, 172, 178, 185, 186, 199, 200, 202, 208
Microsoft, 30, 190
minimum wage, 155
mining, 96, 106, 107, 108, 109, 113, 114, 115, 116
Ministry of Education, 195
minority, 32, 34, 149, 151, 152
mixing, 122, 196
mobile device, 205
mobile phone, 3, 4, 123, 200, 202, 206, 207, 211, 212
mobility, x, 117, 118
model, 11, 18, 20, 32, 50, 68, 109, 110, 111, 135, 152, 162, 163, 187, 200, 218
modeling, 15, 16
models, 14, 15, 18, 20, 48, 96, 110, 111, 115, 156
money, 32, 79, 124, 125, 154
morbidity, 166
mortality, 132, 166
mortality risk, 132
motivation, 16, 50, 57, 79, 128, 143, 144, 204, 210, 212
multiculturalism, 161
multimedia, 24, 25, 73, 98, 115, 184, 187, 202, 203, 216
multiple sclerosis, 185, 187
music, 3, 72, 85
MySpace, ix, 71, 72, 73, 77, 79, 80, 82, 169

N

narratives, xi, 178, 184, 187
national borders, 166
National Center for Education Statistics, 103
national security, 32, 149
Native Americans, 154
negative influences, x, 118
negativity, 163
negotiation, 21
network, 28, 32, 35, 36, 37, 38, 39, 40, 48, 51, 75, 113, 119, 125, 129, 133, 144, 150, 159, 164, 168, 190, 194, 201, 203, 204, 205
networking, 36, 43, 72, 73, 74, 75, 76, 79, 80, 169
neuroscience, 65, 200
new media, 63, 70, 79, 218
newspapers, 78, 168
nodes, 35, 36, 38, 39

O

obesity, 62, 132, 144
objectives, 19, 113
objectivity, 206
observations, 4, 22, 56, 119, 145
older adults, 119, 128, 219
older people, x, 117, 118, 119, 120, 121, 122, 126, 127, 129, 135, 219
online information, 96, 141
online learning, 97, 115
online media, 150, 159
openness, 68, 152, 160
optimization, 116

order, viii, xi, 1, 20, 27, 29, 30, 32, 42, 43, 44, 45, 57, 60, 61, 64, 79, 86, 87, 90, 96, 108, 111, 121, 142, 147, 156, 159, 168, 174, 205, 206, 216
outsourcing, 79
overlay, 108
overweight, 135, 144
ownership, xii, 8, 23, 185, 215

P

parameters, 29, 111
parental consent, 179
parents, 125, 126
participatory democracy, 163
passive, 121
password, 124
path analysis, 162
pathways, 118
pedagogy, 22, 24, 67
peer support, 92
perceptions, 14, 18, 20, 50, 119, 149, 151, 162, 181, 207, 208, 210, 211, 212
performers, 113
perseverance, 16
personal computers, 178
personal identity, 20
personal life, 57, 65, 67
personal relations, 65
personal relationship, 65
personality, 78, 84
persuasion, 187
photographs, 85
physical activity, 179, 184, 185
physical environment, 204
physics, 23, 59, 64
pilot study, ix, 95, 106, 108, 109, 110, 113
planning, 3, 8, 10, 77, 165, 174, 175
polarization, viii, x, 28, 37, 45, 147, 148, 161, 172
policy makers, 33, 96, 169
policy responses, 167
political participation, 152, 154, 156, 157, 158, 159, 164, 216
political parties, 50
politics, 24, 31, 32, 34, 49, 55, 56, 61, 64, 66, 67, 141, 150, 153, 154, 158, 161, 162, 163
poor readers, 28
population, x, 73, 110, 118, 131, 135, 138, 139, 142, 144, 149, 154, 168, 174, 178, 184, 185, 186, 216
population size, 135
portfolios, 74
positive relation, x, 131, 135, 136, 141
poverty, 118, 129, 219
power, 19, 28, 30, 33, 49, 80, 88, 96, 110, 153
power relations, 80
predictors, 135, 137, 140, 141, 151, 152, 159
preference, 185, 186
preparedness, 174
presidential campaigns, 32
pressure, 60
prevention, 179
primary data, 106
prior knowledge, 2, 13
privacy, 114
probability, 110, 111
probability distribution, 110
probe, 9, 21, 180
problem solving, 2
production, 11, 91, 121, 210
productivity, xii, 68, 189, 199
profanity, 188
professional development, 96, 98, 99
profit, xi, 30, 75, 177
program, x, 3, 96, 98, 124, 131, 138, 139, 140, 196, 203, 219
psychology, 65
public domain, 167
public health, 66, 144, 166, 167, 169, 170, 173, 175
public opinion, vii, 27, 31, 33
public policy, 48, 219
public safety, 149
public support, 149
publishers, 195

Q

qualitative research, 206
quality of life, 118, 215, 218
questioning, 17, 181
quizzes, 207

R

race, 154, 156, 182
radar, 50
radio, 56, 69, 200
rainforest, 11
range, viii, 12, 14, 16, 20, 22, 53, 54, 66, 85, 111, 113, 148, 149, 178, 180, 205
ratings, 154, 179, 188
readership, 34, 55, 58, 59
reading, ix, 5, 15, 22, 56, 83, 124, 136, 142, 200, 205, 210, 212
real time, 74, 200, 204
reality, 15, 18, 32, 115, 161, 178, 187
reason, 9, 57, 59, 170, 174

reasoning, xii, 199, 201
recognition, 60
reconstruction, 14
recruiting, 76, 77, 167
referees, 79
refining, 8
reflection, 2, 10, 14, 24, 28
reflective practice, 24
regional economies, 219
regression, x, 108, 131, 135, 137, 140, 145, 156
regression analysis, x, 131, 135, 156
rehabilitation, xii, 215
reinforcement, 119
relationship, x, 4, 9, 11, 13, 17, 20, 35, 103, 119, 135, 147, 148, 150, 152, 159, 160, 184, 204, 216, 218
relevance, 2, 14, 194, 204, 207, 208, 209, 210
reliability, 73, 87
religion, 55
religiosity, 149
reputation, 57
residuals, 111
resources, vii, xi, xii, 15, 21, 32, 56, 74, 75, 85, 96, 97, 98, 99, 103, 104, 106, 107, 108, 109, 111, 112, 113, 114, 115, 116, 119, 120, 127, 132, 133, 142, 165, 168, 174, 175, 190, 196, 205, 213, 215
restructuring, 2
returns, 190, 194
rhetoric, 6, 36, 67
risk, 118, 121, 127, 132, 142
risk factors, 132
role-playing, 61, 77
room temperature, 12
rural population, 184, 185

S

safety, 120, 151, 166, 204
sales, 108, 178
sampling, 135
sanctions, 173
satisfaction, 12, 57
scaling, 37
scholarship, 57, 59, 60
school, xi, 10, 12, 13, 16, 17, 23, 29, 76, 77, 81, 96, 99, 101, 102, 103, 104, 105, 152, 166, 171, 177, 179, 180, 182, 183, 184, 185, 186, 205, 206, 207, 208, 211, 212
school work, 208
schooling, 15, 139
scientific knowledge, 23
scores, 112
search, xii, 7, 21, 36, 49, 75, 86, 88, 89, 92, 97, 98, 99, 100, 103, 105, 133, 141, 143, 153, 189, 190, 191, 193, 194, 195, 197, 202, 203
search terms, 100, 141
searches, 88, 99, 190, 192, 194
searching, xii, 3, 86, 88, 89, 141, 179, 189, 190, 193, 195
security, 151, 162
seed, 68
selectivity, 162
self-confidence, 118
self-efficacy, 119
self-empowerment, 28
self-esteem, 126
self-expression, 28, 84
self-organization, viii, 27, 47
self-presentation, 52, 57
semantics, 107
Senate, 32
sensation, ix, 71, 72
September 11 (9/11), 48, 51, 149, 151, 162, 164
SES, x, 131, 132, 133, 134, 136, 141, 142, 145
sex, 173, 188
sexuality, 186
shade, 103, 104, 105
shape, 4, 7, 31, 32, 107, 122
shaping, x, 118, 126, 200, 217
shares, 30, 123, 136
sharing, xii, 2, 3, 4, 7, 8, 9, 10, 14, 15, 18, 21, 22, 28, 30, 50, 56, 59, 67, 68, 72, 73, 76, 85, 90, 133, 195, 199, 202, 203, 211, 212
siblings, x, 118, 126
signs, 11
skewness, 28, 37
skills, vii, xi, xii, 1, 13, 15, 16, 29, 60, 74, 79, 100, 141, 143, 145, 152, 177, 199, 201, 203, 204, 206, 211, 212, 216
smokers, x, 131, 134, 138, 139, 140, 143
smoking, x, 131, 134, 138, 139, 140, 141, 142
smoking cessation, x, 131, 134, 138, 139, 140, 141, 142
SMS, vi, 3, 4, 82, 199, 200, 202, 206, 207
SNS, 63, 70
social activities, 135
social capital, x, 118, 121, 127, 128, 131, 132, 133, 134, 135, 136, 137, 141, 142, 143, 144, 164, 219
social change, 30, 121, 125
social class, 118
social construct, 2, 19, 218
social distance, 67
social environment, 215
social exclusion, 118, 119, 128, 129, 219
social influence, 50

social institutions, 166
social justice, 169, 175
social learning, 30
social life, 24, 143
social movements, 161
social network, 28, 33, 35, 36, 37, 39, 47, 48, 52, 72, 73, 74, 75, 76, 77, 78, 79, 80, 81, 82, 90, 119, 120, 129, 132, 133, 163, 169, 170, 199, 200, 201, 202, 203, 204, 205
social participation, 144, 215
social phenomena, 35
social relations, 35, 132
social relationships, 132
social skills, 142
social status, 132
social support, 134, 145, 215
socialization, 122
socioeconomic status, xi, 147
software, 34, 49, 50, 54, 85, 92, 109, 135, 171, 190, 192, 216
space, 4, 6, 7, 8, 9, 10, 16, 18, 22, 29, 36, 49, 64, 103, 122, 123, 127, 172, 194, 204, 218
spam, 78, 84, 107
spectrum, 152, 216, 218
speculation, 44
speech, 5, 6, 9, 10, 151, 152, 155, 161, 163, 216
speed, 132, 138, 187, 200, 205, 212
SPSS, 109, 135, 180
standard deviation, 37, 38, 101
standardization, 100
standards, 31, 61, 80, 195, 216
statistics, 37, 73, 110, 180
stereotypes, 186
stock, 141
storage, 194, 207
strategies, ix, 3, 95, 96, 101, 166, 167, 174, 185
strength, 15, 20, 30, 45, 47, 113, 156, 158, 161
supply, 15, 172
Supreme Court, 155
survival, xii, 167, 178, 215
synthesis, 218

T

target populations, 179, 185
targets, 85, 97
taxonomy, 85
teachers, 3, 13, 20, 23, 55, 96, 97, 98, 99, 100, 103, 106, 107, 109, 110, 113, 115, 202, 204, 205, 207, 214
teaching, ix, 2, 3, 4, 10, 11, 13, 18, 20, 22, 25, 67, 71, 74, 75, 97, 99, 100, 115, 124, 125, 196, 201, 202, 204, 206, 212, 213, 214
teaching experience, 97, 100
technician, 122
technological developments, 170
teenagers, 184, 202
teens, 73, 82, 178, 185, 187, 202
telephone, 124
television, 168, 172
tension, 60, 124
terrorism, 149, 162
thinking, xi, 4, 5, 6, 7, 8, 9, 10, 11, 12, 13, 14, 17, 18, 22, 53, 56, 178, 212
thoughts, xii, 2, 5, 6, 9, 11, 20, 22, 29, 30, 65, 69, 73, 78, 87, 89, 93, 199, 200, 201, 207, 208
threat, 31, 114, 149, 151, 153, 187
time periods, 101, 104
time use, 102, 109
tourism, 49, 52
tracking, 20, 100, 101, 106
traditional gender role, 182
traffic, 28, 34, 96, 100, 102, 103, 105, 107, 194
training, 31, 62, 75, 118, 122, 126, 127, 201, 204, 211, 218
traits, 182
transactions, 100
transcription, 20
transformation, 106, 108, 113
transformations, 52, 93
translation, 64
transmission, 60, 61
transportation, 178, 187
traumatic brain injury, 218
Trent Lott, 31
trial, x, 16, 117, 122, 138, 179, 187, 219
trust, x, 30, 118, 119, 131, 134, 135, 136, 138, 140, 141, 149, 164

U

UNESCO, 134, 145
universe, ix, 71, 72, 195
universities, 64, 76, 77, 78, 80, 81, 82, 102, 155, 204, 205
user data, 113

V

variables, x, xi, 63, 64, 65, 107, 108, 109, 110, 111, 113, 131, 134, 135, 136, 137, 140, 142, 143, 154, 156, 157, 158, 159, 164, 177
variance, 38, 111, 112, 135, 156, 157, 158
vector, 107, 190, 192, 194
video games, xi, 79, 177, 178, 183, 185, 186

violence, xi, 177, 184, 186
visions, 1, 2, 19, 20, 114, 218
voice, 6, 13, 17, 19, 31, 32, 161, 182
voice-overs, 182
voicing, 31
voting, 31, 74, 154
voting record, 154
vulnerability, 8
Vygotsky, 6, 14, 22, 25

W

wealth, 91, 151, 170
web, vii, ix, 5, 15, 27, 29, 35, 48, 49, 50, 72, 84, 87, 88, 93, 95, 96, 97, 98, 99, 100, 101, 102, 106, 109, 112, 113, 114, 115, 116, 129, 171, 175, 196, 200, 201, 217, 218, 220
web pages, vii, 27, 29, 35, 97, 100, 113
weblog, viii, 29, 32, 33, 34, 49, 51, 53, 54, 63, 65, 67, 70, 92
websites, ix, 15, 29, 34, 71, 124, 141, 160, 162, 169, 170, 179, 216
welfare, 138, 172
women, 69, 121, 122, 123, 133, 139, 141, 153, 160, 173
working groups, 2
workload, 67
workplace, 51, 63, 91, 92, 163, 166
World Wide Web, 72, 75, 134, 144
writing, ix, xii, 2, 4, 5, 9, 15, 21, 22, 54, 57, 59, 60, 62, 64, 65, 66, 67, 79, 83, 86, 87, 89, 142, 199, 200, 201, 204, 206, 207, 208, 209, 210, 211, 212, 214
writing process, 9, 21, 210, 212

X

XML, 190, 191, 192, 194, 196, 197